INTEGRAL GEOMETRY AND CONVEXITY

Proceedings of the International Conference

INTEGRAL GEOMETRY AND CONVEXITY

Wuhan, China 18 – 23 October 2004

Editors

Eric L Grinberg
University of New Hampshire, USA

Shougui Li
Wuhan University of Sciences and Technology, China

Gaoyong Zhang
Polytechnic University, USA

Jiazu Zhou
Guizhou Normal University, China

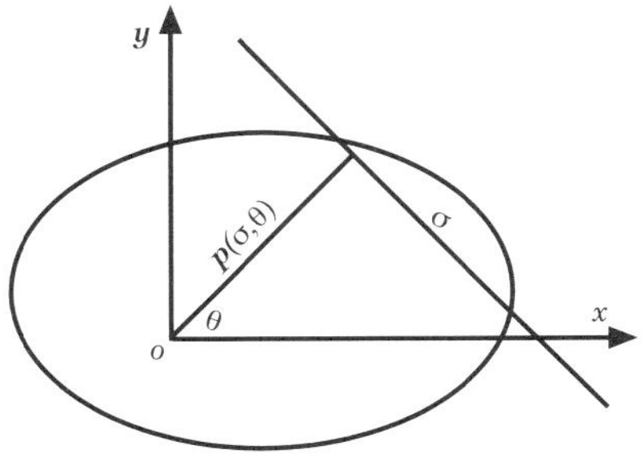

World Scientific

NEW JERSEY • LONDON • SINGAPORE • BEIJING • SHANGHAI • HONG KONG • TAIPEI • CHENNAI

Published by

World Scientific Publishing Co. Pte. Ltd.
5 Toh Tuck Link, Singapore 596224
USA office: 27 Warren Street, Suite 401-402, Hackensack, NJ 07601
UK office: 57 Shelton Street, Covent Garden, London WC2H 9HE

British Library Cataloguing-in-Publication Data
A catalogue record for this book is available from the British Library.

INTEGRAL GEOMETRY AND CONVEXITY
Proceedings of the International Conference

ISBN 981-256-513-2

Printed in Singapore by World Scientific Printers (S) Pte Ltd

PREFACE

Integral geometry, known as geometric probability in the past, originated from Buffon's needle experiment. Crofton, Poincare, Blashke, Chern, Santalo and others made significant contributions in the past centuries. The applications of this field vary from the medical sciences to other mathematic branches including algebra, geometric inequalities, differential equations, topology, and geometric convexity. Remarkable advances have also been made in several areas that involve the theory of convex bodies. In 2002, a select group of mathematicians believed it was the time to convene the first conference on these important fields in China.

The result was the First International Conference on Integral Geometry and Convexity Related Topics in China, which was held at Wuhan University of Science and Technology from October 18 to 23, 2004 and sponsored by the Mathematical Associations of Hubei Province and Wuhan City. The organizers carefully selected international researchers who are leaders in their specialties to discuss their recent results and their ideas on the trends of future research. The only regret is that some known mathematicians could not present due to their own schedule.

The program consisted of talks on integral geometry, convex geometry, complex geometry, probability, statistics and other convexity-related branches. The principal speakers were Rolf Schneider, Eric Grinberg, and Ralph Howard; they were joined by ten other leading international mathematicians and eight domestic Chinese researchers. The conference was purposely designed to facilitate the discussion and exchange of ideas among researchers with various specialties and we believe that objective was achieved.

It is especially gratifying to have the talks, which were presented at the Conference collected together in these Proceedings (World Scientific Publishers, Singapore). The major themes include probabilistic and analytic methods in the study of convex bodies, especially in high dimensions, applications of integral geometry and convex-geometric methods to other branches of mathematics, isoperimetric-type inequalities, radon transforms, and applications to medical, economic and information sciences. The editors of the Proceedings are Jiazu Zhou, Gaoyong Zhang, Eric Grinberg and Shougui Li, all of whom should also be recognized for their hard work and diligence in the planning and execution of all aspects of this international conference. Since I was unable to attend the conference due to physical mobility difficulties, I am looking forward to the publication of the Proceedings.

On behalf of the organizing committee, we would also like to express our sincere gratitude to the invited speakers and audience of over 50 research mathematicians whose participation made this conference a great success. We hope that this conference is the start of many more such gatherings.

Chuan Chih Hsiung
Lehigh University
June, 2005

FORWARD

For several decades students in Wuhan were introduced to and mentored in the subjects of integral geometry, convexity, geometric inequalities, geometric probability and allied fields under the direction of Ren Delin and with the support of S.S. Chern and C.C. Hsiung. Indeed, this mathematical locale provided unique opportunities for students to learn about these beautiful and important subjects, which might otherwise have been quite inaccessible to these future mathematicians. Wuhan graduates went on to make broad and substantial research contributions. Thus it seemed particularly appropriate to hold a conference on these topics in Wuhan City and Hubei Province, a special place for integral and convex geometry, and in a special year, when Professor Ren Delin was about to retire at age 70. Some participants, myself included, were fortunate to have visited Wuhan in earlier decades and hence to witness the vast and rapid development of academic facilities and growth of support of the mathematical sciences. The conference infrastructure was superb. The local organizers are sincerely thanked for the outstanding environment which they crafted for the event and for the great care they took each day to insure that all had what was needed. We also thank the National Science Foundation which supported some of the organizers through USA-China Cooperative grant DMS-9906856.

The community of participants was lively, with broad representation which made a stimulating atmosphere for the exchange of ideas, and which is reflected in the proceedings. All regretted that medical travel restrictions prevented S.S. Chern from attending the event. However, Professor Chern sent a letter of support to the conferees which was read during the opening ceremonies, and his support was felt throughout. It was most saddening to learn of his untimely death in December, 2004. At the same time, it is fortunate that the conference was held during Professor Chern's lifetime. For many conferees it provided the last link to a great mathematician and a great man.

Eric L. Grinberg
University of New Hampshire

CONTENTS

VOLUME INEQUALITIES FOR SETS ASSOCIATED WITH CONVEX BODIES

STEFANO CAMPI
Dipartimento di Matematica Pura e Applicata "G. Vitali"
Università degli Studi di Modena e Reggio Emilia
Via Campi 213/B 41100 Modena, Italy
E-mail: campi@unimo.it

PAOLO GRONCHI
Istituto per le Applicazioni del Calcolo
Sezione di Firenze
Via Madonna del Piano
CNR Edificio F, 50019
Sesto Fiorentino (FI), Italy
E-mail: paolo@fi.iac.cnr.it

This paper deals with inequalities for the volume of a convex body and the volume of the projection body, the L^p-centroid body, and their polars. Examples are the Blaschke-Santaló inequality, the Petty and Zhang projection inequalities, the Busemann-Petty inequality. Other inequalities of the same type are still at the stage of conjectures. The use of special continuous movements of convex bodies provides a general approach to this subject. A family of inequalities, depending on a parameter $p \geq 1$ and proved by Lutwak for $p = 1$ and $p = 2$, is obtained.

1. Introduction and preliminaries

This paper is devoted to some classical inequalities of Convex Geometry involving the volume of an n-dimensional convex body and the volume of a further body associated to the given one. More precisely, our attention is focused on the projection body, the L^p-centroid body and their polar bodies.

Our approach comes from the idea that the most part of results connected with these inequalities can be deduced by the same general method, which is based on the use of special continuous movements of the bodies we are dealing with.

Let K be a convex body in $\mathbb{R}^n$, that is a n-dimensional compact convex set, and assume that the origin is an interior point of K.

The support function of the convex body K is defined as

$$h_K(u) = \max_{x \in K} \langle u, x \rangle, \text{ for } u \in \mathbb{R}^n,$$

where $\langle \cdot\, , \cdot \rangle$ is the usual scalar product in $\mathbb{R}^n$, and the radial function of K as

$$\rho_K(u) = \max\{r \in \mathbb{R} : ru \in K\}, \text{ for } u \in \mathbb{R}^n.$$

The n-dimensional volume $V(K)$ of K can be expressed in terms of the radial function by

$$V(K) = \frac{1}{n} \int_{S^{n-1}} \rho_K^n(z)\, dz,$$

where S^{n-1} is the unit sphere in $\mathbb{R}^n$.

The *polar body* K^* of K can be defined as

$$K^* = \{x \in \mathbb{R}^n | \langle x, y \rangle \leq 1, \forall y \in K\}.$$

The polar body of K depends on the location of the origin. It is easy to see that $(K^*)^* = K$ and that

$$\rho_{K^*}(u) = \frac{1}{h_K(u)}, \text{ for } u \in \mathbb{R}^n.$$

The *projection body* of K is the convex body ΠK such that

$$h_{\Pi K}(u) = \frac{1}{2} \int_{\partial K} |\langle u, v \rangle|\, dv, \text{ for } u \in \mathbb{R}^n,$$

where dv denotes the area element at the point on ∂K whose outer unit normal is v. It is clear from the definition that $h_{\Pi K}(u)$ is the $(n-1)$-dimensional volume of the projection of K orthogonal to u.

For every Borel subset ω of S^{n-1}, we define the *area measure* $\sigma_K(\omega)$ of K as the $(n-1)$-dimensional Hausdorff measure of the reverse image of ω through the Gauss map. Recall that the Gauss map sends each point on ∂K to the set of outward unit normal vectors to ∂K at that point. Therefore, the support function of ΠK can be rewritten as

$$h_{\Pi K}(u) = \frac{1}{2} \int_{S^{n-1}} |\langle u, v \rangle|\, d\sigma_K(v), \text{ for } u \in \mathbb{R}^n.$$

If ∂K is sufficiently smooth and its Gauss curvature is strictly positive, then

$$h_{\Pi K}(u) = \frac{1}{2} \int_{S^{n-1}} \frac{|\langle u, v \rangle|}{\gamma_K(v)}\, dv, \text{ for } u \in \mathbb{R}^n,$$

being $\gamma_K(v)$ the Gauss curvature of ∂K at the point where the outward unit normal vector is v. Such a representation shows that $h_{\Pi K}$ is the *cosine transform* of $\frac{1}{\gamma_K}$.

Each projection body is a *zonoid*, namely it is the limit, in the Hausdorff metric, of a sum of segments. Here by sum it is meant the Minkowski addition of subsets of $\mathbb{R}^n$

$$A + B = \{x \in \mathbb{R}^n : x = a + b, a \in A, b \in B\}.$$

Zonoids play an important role in Convex Geometry, as well as in different areas of Mathematics. In particular, every full-dimensional zonoid turns out to be the projection body of a convex body.

Given a convex body K, for each real number $p \geq 1$, the L^p*-centroid body* $\Gamma_p K$ of K is the convex body whose support function is

$$h_{\Gamma_p K}(u) = \left\{ \frac{1}{c_{n,p} V(K)} \int_K |\langle u, z \rangle|^p \, dz \right\}^{\frac{1}{p}}, \text{ for } u \in \mathbb{R}^n,$$

where

$$c_{n,p} = \frac{\kappa_{n+p}}{\pi \kappa_n \kappa_{p-1}},$$

and

$$\kappa_r = \pi^{\frac{r}{2}} / \Gamma(1 + \frac{r}{2}).$$

Notice that κ_n is the volume of the unit ball B^n of $\mathbb{R}^n$ and the constant $c_{n,p}$ is such that $\Gamma_p B^n = B^n$.

The above definition can be extended to compact bodies in $\mathbb{R}^n$.

Up to constants, $\Gamma_1 K$ is known in the literature as the *centroid body* ΓK of K and $\Gamma_2 K$ as the *Legendre ellipsoid* of K. If K is an origin-symmetric convex body, it turns out that the boundary of ΓK is the locus of the centroids of all the halves of K obtained by cutting K with hyperplanes through the origin.

By using polar coordinates in integration, one has that

$$h_{\Gamma K}(u) = \frac{1}{n c_{n,1} V(K)} \int_{S^{n-1}} |\langle u, z \rangle| \rho_K^{n+1}(z) \, dz,$$

which shows that the centroid body ΓK is, up to a constant, the projection body of ΛK, the *curvature image* of K, i.e. the convex body such that $\gamma_{\Lambda K}(z) = \rho_K^{-n-1}(z)$, for every $z \in S^{n-1}$. The Minkowski Theorem guarantees the existence of such a body (see, for example, [32], Ch. 7.1).

If $\Gamma_\infty K$ is interpreted as a limit of (1), as $p \to \infty$, then

$$\Gamma_\infty K = conv(K \cup (-K)) ,$$

where *conv* stands for the convex hull.

For further details related to the content of this preliminary section, we refer to the books of Gardner [11] and Schneider [32] and to the articles by Goodey and Weil [12], Lutwak [17], Lutwak and Zhang [19], Milman and Pajor [23].

As a final remark, we notice that there are other bodies that can be associated to a given one: the *intersection body*, the *cross-section body*, the *Blaschke body*, etc. As Richard Gardner writes in the Introduction of his book [11], *geometric tomography houses a zoo of strange geometric bodies, powerful integral transforms, and exotic but highly effective inequalities.*

2. Main volume inequalities

In this section we list the main inequalities which involve the volume of the bodies introduced in Section 1. At present, some of them are still conjectures. Further information and details on inequalities of this type can be found in [17].

2.1. *The Blaschke-Santaló inequality*

For every convex body K, assume that the origin is chosen so that $V(K^*)$ is minimum. The Blaschke-Santaló inequality states that

$$V(K)V(K^*) \le \kappa_n^2 , \tag{1}$$

where equality holds if and only if K is an origin-symmetric ellipsoid. The quantity on the left-hand side of (1) is called *volume product* of K. Inequality (1) was proved for $n \le 3$ by Blaschke [1], [2] and for all n by Santaló [31]. The equality conditions were proved by Saint Raymond [30] in the symmetric case and by Petty [26] in the general case.

2.2. *Mahler's conjecture*

It was conjectured by Mahler [20] that the minimum of the volume product is attained when K is a simplex, that is

$$V(K)V(K^*) \ge \frac{(n+1)^{n+1}}{(n!)^2} . \tag{2}$$

Mahler [21] proved that (2) holds if $n = 2$ and Meyer [22] that in this case equality occurs only for triangles.

For origin-symmetric convex bodies it has been conjectured that n-parallelotopes (and their polars, i.e. cross-polytopes) minimize the volume product, hence

$$V(K)V(K^*) \geq \frac{4^n}{n!}\,. \tag{3}$$

For $n = 2$, inequality (3) was proved by Mahler [21] and Reisner [28] showed that parallelograms are the only minimizers. For $n > 3$ there are bodies, different than parallelotopes and their polars, for which (3) is an equality, as shown by Saint Raymond [30]. Reisner [27], [28] and Saint Raymond [30] proved (3) for special classes of convex bodies, namely for zonoids and for the affine images of convex sets symmetric with respect to the coordinate hyperplanes. A simpler proof of Reisner's result was given in [13].

Bourgain and Milman [4] proved that there exists a constant c, not depending on the dimension, such that

$$V(K)V(K^*) \geq c^n \kappa_n^2\,.$$

2.3. *The projection body conjectures*

The problem of finding minimizers and maximizers of the functional $V(K)^{1-n}V(\Pi K)$ in the class of all convex bodies is open.

In 1972 Petty [25] conjectured that ellipsoids are the only minimizers.

As far as the maximum of $V(K)^{1-n}V(\Pi K)$ is concerned, Brannen [5] conjectured that, in the class of all convex bodies, simplices are maximizers. The same author conjectured that the largest centrally symmetric subset of a simplex gives a sharp upper bound of the functional in the class of centrally symmetric convex bodies.

2.4. *The Petty projection inequality*

Petty [25] proved that

$$V(K)^{n-1}V(\Pi^* K) \leq \left(\frac{\kappa_n}{\kappa_{n-1}}\right)^n, \tag{4}$$

with equality if and only if K is an ellipsoid. Notice that Petty's conjecture in 2.3 and the Blaschke-Santaló inequality (1) imply (4).

2.5. *The Zhang projection inequality*

The reverse inequality of (4) is due to Zhang [34]:

$$V(K)^{n-1}V(\Pi^*K) \geq \binom{2n}{n} n^{-n}, \tag{5}$$

with equality if and only if K is a simplex.

2.6. *The L^p-Busemann-Petty inequality*

Let $p \geq 1$. Then

$$V(\Gamma_p K)V(K)^{-1} \geq 1, \tag{6}$$

with equality if and only if K is an origin-symmetric ellipsoid. For $p = 1$, inequality (6) was proved by Petty [24] via the Busemann random simplex inequality (see [11], Theorem 9.2.6). For $p = 2$, Blaschke [3] proved (6) when $n = 3$. For $p = 2$ and general n, (6) was proved by John [14]. For general p, inequality (6) was proved by Lutwak, Yang and Zhang [18] and, in a different way, by Campi and Gronchi [8].

Concerning a reverse version of (6), let us notice that the functional we are considering is not bounded from above in the class of all convex bodies. A natural assumption is to restrict ourselves to the bodies containing the origin. It has been conjectured that in such a class simplices with one vertex at the origin provide the maximum. In [9] this conjecture is proved for $n = 2$ (see also Section 4).

2.7. *The L^p-Blaschke-Santaló inequality*

Let $p \geq 1$. Then

$$V(\Gamma_p^* K)V(K) \leq \kappa_n^2, \tag{7}$$

with equality if and only if K is an origin-symmetric ellipsoid, as proved by Lutwak and Zhang in [19]. The name of this inequality comes from the fact that if K is origin-symmetric and p tends to infinity, then (7) gives (1).

Even in this case, it has been conjectured that suitable simplices are minimizers of the left-hand side of (7) in the class of all convex bodies containing the origin. A result in the two-dimensional case is contained in Subsection 4.3.

3. The Rogers and Shephard method

Most of the results described in the previous section can be obtained by the use of *shadow systems*, which were introduced by Rogers and Shephard in [29] and [33].

A *shadow system* along the direction v is a family of convex bodies $K_t \subset \mathbb{R}^n$ that can be defined by

$$K_t = conv\{z + \alpha(z)t\, v : z \in A \subset \mathbb{R}^n\}\,, \tag{8}$$

where A is an arbitrary bounded set of points in $\mathbb{R}^n$, α is a real bounded function on A, and t runs in an interval of the real axis. The function α in (8) is called the *speed function* of the shadow system.

As proved in [29], *the volume of K_t is a convex function of t.* The proof is based on the fact that the length of each chord of K_t parallel to v turns out to be a convex function of t. This convexity result was extended by Shephard [33] to mixed volumes of shadow systems along the same direction v.

We recall that the mixed volume of the convex bodies K_1, K_2, ..., K_n can be defined by

$$V(K_1, K_2, \ldots, K_n) = \frac{1}{n!}\sum_{j=1}^{n}(-1)^{n+j} \sum_{i_1<\cdots<i_j} V(K_{i_1}+K_{i_2}+\cdots+K_{i_j})\,. \tag{9}$$

As proved by Minkowski (see [32], Ch. 5), mixed volumes are the coefficients of the homogeneous polynomial $V(t_1K_1 + \cdots + t_nK_n)$ of degree n in the variables t_1, t_2, ..., t_n. Special instances of mixed volumes are the *quermassintegrals* $W_i(K)$ of a convex set K, which are defined by (9), for $i = 0, 1, \ldots, n$, by taking $(n - i)$ copies of K and i copies of the unit ball B^n. Notice that every mixed volume can be expressed by an integral. An example that we shall use later is given by

$$V(K, \ldots, K, L) = V_1(K, L) = \frac{1}{n}\int_{S^{n-1}} h_L(v)\, d\sigma_K(v). \tag{10}$$

Shephard's proof of the convexity of mixed volumes under shadow systems is based on the remark that every shadow system K_t along v can be seen as the projection of an $(n+1)$-dimensional convex body onto a fixed hyperplane orthogonal to w with respect to the direction $w - tv$.

A special type of shadow system is naturally related to the well known Steiner process of symmetrization. Precisely, fix a direction v and let

$$K = \{x + yv \in \mathbb{R}^n : x \in K|v^{\perp}, y \in \mathbb{R}, f(x) \leq y \leq g(x)\}\,;$$

here $K|v^\perp$ denotes the orthogonal projection of K onto the hyperplane $v^\perp$ through the origin, orthogonal to v, and f and $-g$ are convex functions on $K|v^\perp$. The shadow system with speed function $\alpha(x) = -(f(x) + g(x))$ and $t \in [0,1]$ is such that $K_0 = K$, $K_1 = K^v$, the reflection of K in the hyperplane $v^\perp$, and $K_{1/2}$ is the Steiner symmetral of K with respect to $v^\perp$. In this process, the volume is trivially constant, while the behavior of the other quermassintegrals is described by Shephard's result on mixed volumes.

Thus the Steiner symmetral of a convex body K can be obtained through a shadow system along a direction v in which the speed function is constant on each chord parallel to v. Shadow systems of this type are called *parallel chord movements.* So, a parallel chord movement along the direction v is a family of convex bodies K_t in $\mathbb{R}^n$ defined by

$$K_t = \{z + \alpha(x)t\, v : z \in K, x = z - \langle z, v\rangle v\},$$

where α is a continuous real function on $v^\perp$. Notice that the above definition requires that α must be chosen so that, for every t, the set K_t is convex.

If the speed function α is an affine function (that is, $\alpha(x) = \langle x, u\rangle + k$, for some vector u and real constant k), then it is easy to see that K_t is an affine image of K, for every t in the range of the movement.

Shadow systems can be applied successfully according to the following Shephard argument (see [33]): *If a functional defined in the class of all convex sets is continuous, invariant under reflections and convex with respect to the parameter t of any parallel chord movement, then it attains its minimum at the ball among all sets of given volume.* Here the continuity refers to the Hausdorff metric.

This statement follows from the well-known property of the Steiner symmetrization, that, if suitably repeated, changes every convex set in a ball.

Consequences of this procedure are, for example, classical isoperimetric type inequalities for quermassintegrals (see [6], p. 144, [11], p. 372). Other geometric functionals have the same convex behavior under shadow systems; see, for example, [33], [7] and the next section.

A question is whether the same convexity property of the functionals can help in finding reverse inequalities.

We are able to give a positive answer only in the case $n = 2$. In order to do this, it is convenient to introduce the class Ω_n of n-dimensional convex bodies K with the following property: *If K_t is a parallel chord movement, $t \in [-1,1]$, and $K = K_0$, then the speed of the movement is an affine function.*

Assume now that $F(K)$ is a functional defined in the class of all n-dimensional convex bodies and suppose F has a maximum in that class. If F is convex under parallel chord movements, then a maximizer of F must belong to Ω_n. Notice that, if F is bounded from above and convex under parallel chord movements, then F must be invariant under affine transformations. Moreover, if F is *strictly* convex, i.e. $F(K_t)$ is strictly convex unless the speed of K_t is affine, then all maximizers of F belong to Ω_n. As shown in [7], all simplices are in Ω_n, triangles are the only polygons in Ω_2, and, for $n \geq 2$, in Ω_n there are other bodies than simplices. For $n = 2$, by an approximation argument, we can conclude that triangles are maximizers of all continuous functionals which are convex under parallel chord movements (provided the maximum exists).

For $n = 2$, if one considers only centrally symmetric convex sets, then parallelograms play the same role as triangles in the general case.

The same method can be applied also to *linear* invariant functionals which are not bounded from above. Since these are the cases we shall deal with in the next section, we give here all the details.

A linear invariant continuous functional is bounded in the class of bodies containing the origin. Indeed, by John's theorem ([15], Theorem III), we can restrict ourselves to bodies containing a ball of radius one and contained in a ball of radius n, with the same center.

If F is a convex functional under parallel chord movements, then it is convex under translations, hence every maximizer of F has the origin as an extreme point. In particular, if P is a polytope, we can assume, without decreasing the value of F, that the origin is one of its vertices.

Let F be a functional defined in the class of all convex bodies containing the origin, which is continuous with respect to the Hausdorff metric, linear invariant and convex under parallel chord movements. Let P be a polygon and let $0, v_1, v_2, \ldots, v_m$ be its vertices clockwise ordered.

Let us consider the shadow system $\{P_t : t \in [t_0, t_1]\}$, $t_0 < 0 < t_1$, along v_2, with speed 1 at v_1 and 0 at the other vertices. If t_0 and t_1 are sufficiently close to 0, then only the triangle $0v_1v_2$ moves, while the remaining part of P keeps still. Let us choose $[t_0, t_1]$ as the largest interval such that the area of P_t is constant for all $t \in [t_0, t_1]$. Hence, $\{P_t : t \in [t_0, t_1]\}$ is just a parallel chord movement and P_{t_0} and P_{t_1} have exactly $m - 1$ vertices. By the convexity of F,

$$F(P) \leq \max\{F(P_{t_0}), F(P_{t_1})\}\,.$$

If $m > 4$, iterations of this argument lead to the conclusion that $F(P) \leq$

$F(T)$, where T is a triangle with a vertex at the origin. The continuity of F implies that T is a maximizer in the whole class of plane convex figures containing the origin.

4. Applications

This section contains some applications of two results involving shadow systems and related ideas.

For $v \in S^{n-1}$, let H_v^+ be the halfspace bounded by $v^\perp$ and containing v.

Theorem 4.1. *If K_t, $0 \leq t \leq 1$, is a shadow system along the direction v, then $V(K_t^* \cap H_v^+)^{-1}$ is a convex function of t.*

The proof (see [10]) is based on a generalized form of the Prekopa-Leindler inequality. Notice that the theorem deals only with one of the halves of K^* cut off by $v^\perp$. Clearly, if K_t is a family of origin-symmetric convex bodies, then $V(K_t^*)^{-1}$ is a convex function of t.

Theorem 4.2. *If K_t, $0 \leq t \leq 1$, is a parallel chord movement along the direction v, then $\Gamma_p K_t$, for $p \geq 1$, is a shadow system along v. Hence, $V(\Gamma_p K)$ is convex under parallel chord movement. Moreover, it is strictly convex unless the speed function is linear.*

For the proof of Theorem 4.2 see [8].

4.1.

By applying Theorem 4.1 to the reciprocal of the volume product $V(K)V(K^*)$, we obtain the Blaschke-Santaló inequality (1) for origin-symmetric convex bodies (without the characterization of ellipsoids as unique maximizers).

For $n = 2$, we get the Mahler inequality (3) (again without characterization).

4.2.

Theorem 4.2 immediately gives the L^p-Busemann-Petty centroid inequality (6), with the characterization of minimizers.

The same theorem, for $n = 2$, implies also a reverse form of the inequality.

Namely, *all triangles with a vertex at the origin maximize* $V(\Gamma_p K)V(K)^{-1}$ *in the class of all plane convex bodies containing the origin.*

4.3.

Combining Theorem 4.1 with Theorem 4.2 provides the L^p-Blaschke-Santaló inequality (7) (without characterization) and the following reverse form for $n = 2$:

In the class of all plane convex bodies K *containing the origin, triangles with a vertex at the origin are minimizers of* $V(\Gamma_p^* K)V(K)$.

If one deals with plane origin-symmetric convex bodies, then centered parallelograms are maximizers and minimizers for the functionals in 4.2 and 4.3, respectively.

4.4. *Double entry functionals*

In this subsection we consider geometric functionals depending on two different convex bodies. We already noticed in Section 3 that mixed volumes are convex along shadow systems; for example, if K_t and L_t, $0 \leq t \leq 1$, are shadow systems along the same direction v, then $V_1(K_t, L_t)$ is a convex function of t.

Let us consider now the functional

$$G_p(K, L) = \frac{V_1(K, \Gamma_p L)}{V(K)^{\frac{n-1}{n}} V(L)^{\frac{1}{n}}}, \tag{11}$$

where $p \geq 1$.

Theorem 4.3. *If* K_t *and* L_t, $0 \leq t \leq 1$, *are parallel chord movements along the same direction* v, *then* $G_p(K_t, L_t)$, *for* $p \geq 1$, *is a convex function of* t.

The theorem follows from Theorem 4.2 and from the above mentioned property of mixed volumes.

Theorem 4.4. *For every pair of convex bodies* K *and* L,

$$G_p(K, L) \geq 1, \tag{12}$$

where equality holds if and only if K *and* L *are homothetic ellipsoids, with* L *origin-symmetric.*

Proof. The theorem is a consequence of the Minkowski inequality (see [32], p. 317) and the L^p-Busemann-Petty centroid inequality (6). The equality conditions for these two inequalities imply the second part of the statement. □

The functional $G_p(K,L)$ is not bounded from above. Indeed, if L is moved far from the origin, then $\Gamma_p L$ becomes larger and larger. Moreover, if K and L approach to a couple of $(n-1)$-dimensional sets lying on non parallel hyperplanes, then the volumes of K and L tend to zero, while $V_1(K,L)$ tends to a positive number. Nevertheless, Theorem 4.3 provides a two-dimensional reverse form of inequality (12) when $K = L$.

Theorem 4.5. *For $n = 2$, if K contains the origin, then $G_p(K,K)$ attains its maximum when K is a triangle with a vertex at the origin.*

Let us show that inequality (12), for $p = 1$, implies the Petty projection inequality (4). We have

$$\begin{aligned}
1 \le \min_L G_1(K,L) &= \min_L \frac{(n+1)\kappa_n \int_{S^{n-1}} \int_L |\langle x,u\rangle|\, dx d\sigma_K(u)}{2n\kappa_{n-1} V(K)^{\frac{n-1}{n}} V(L)^{\frac{n+1}{n}}} \\
&= \min_L \frac{(n+1)\kappa_n \int_L \int_{S^{n-1}} |\langle x,u\rangle|\, d\sigma_K(u) dx}{2n\kappa_{n-1} V(K)^{\frac{n-1}{n}} V(L)^{\frac{n+1}{n}}} \\
&= \min_L \frac{(n+1)\kappa_n \int_L h_{\Pi K}(x)\, dx}{n\kappa_{n-1} V(K)^{\frac{n-1}{n}} V(L)^{\frac{n+1}{n}}} \\
&= \frac{(n+1)\kappa_n \int_{\Pi^* K} h_{\Pi K}(x)\, dx}{n\kappa_{n-1} V(K)^{\frac{n-1}{n}} V(\Pi^* K)^{\frac{n+1}{n}}} = \frac{\kappa_n}{\kappa_{n-1} V(K)^{\frac{n-1}{n}} V(\Pi^* K)^{\frac{1}{n}}},
\end{aligned}$$

where we used the fact that the integral of a support function h_M on a set L of given volume is minimum when L is just a level set of h_M, that is, when L is homothetic to M^*.

Notice that $G_1(K,L)$ can be expressed as

$$G_1(K,L) = \frac{(n+1)\kappa_n \int_{\partial K} \int_L |\langle x,u\rangle|\, dx du}{2n\kappa_{n-1} V(K)^{\frac{n-1}{n}} V(L)^{\frac{n+1}{n}}}.$$

The last expression reminds some functionals studied by Lutwak in [16]. Precisely, he showed that Petty's projection body conjecture (see Subsection 2.3) is equivalent to say that the functional

$$\frac{\int_{\partial K} \int_{\partial L} |\langle x,u\rangle|\, dx du}{V(K)^{\frac{n-1}{n}} V(L)^{\frac{n-1}{n}}}$$

attains its minimum, in the class of all pairs of convex bodies, if and only if K and L are homothetic to origin-symmetric polar reciprocal ellipsoids.

In the same paper Lutwak proved also that, for $p = 1$ and $p = 2$, the functional

$$\frac{\int_K \int_L |\langle x, u\rangle|^p \, dx du}{V(K)^{\frac{n+p}{n}} V(L)^{\frac{n+p}{n}}}$$

attains its minimum if and only if K and L are homothetic to origin-symmetric polar reciprocal ellipsoids.

We are able to extend such a result to the case of an arbitrary $p \geq 1$.

Theorem 4.6. *Let $p \geq 1$, then*

$$\int_K \int_L |\langle x, u\rangle|^p \, dx du \geq \frac{n\kappa_{n+p}}{\pi(n+p)\kappa_{p-1}\kappa_n^{\frac{n+2p}{n}}} V(K)^{\frac{n+p}{n}} V(L)^{\frac{n+p}{n}},$$

with equality if and only if K and L are homothetic to origin-symmetric polar reciprocal ellipsoids.

Proof. By the definition of L^p-centroid body, we have that

$$\frac{\int_K \int_L |\langle x, u\rangle|^p \, dx du}{V(K)^{\frac{n+p}{n}} V(L)^{\frac{n+p}{n}}} = \frac{c_{n,p} V(L)}{V(K)^{\frac{n+p}{n}} V(L)^{\frac{n+p}{n}}} \int_K h^p_{\Gamma_p L}(v) \, dv .$$

Therefore, for any fixed L, the minimum of the functional we are dealing with is attained if and only if K is homothetic to a level set of $h_{\Gamma_p L}$. Since the functional is invariant under rescaling, the minimum is attained when $K = \Gamma_p^* L$. Thus,

$$\frac{\int_K \int_L |\langle x, u\rangle|^p \, dx du}{V(K)^{\frac{n+p}{n}} V(L)^{\frac{n+p}{n}}} \geq \frac{nc_{n,p}}{n+p} [V(L)V(\Gamma_p^* L)]^{-\frac{p}{n}} \geq \frac{n\kappa_{n+p}}{\pi(n+p)\kappa_{p-1}\kappa_n^{\frac{n+2p}{n}}},$$

where we used the L^p-Blaschke-Santaló inequality (7).

The above argument provides also the equality conditions. □

References

1. W. Blaschke, Über affine Geometrie VII: Neue Extremeigenschaften von Ellipse und Ellipsoid, *Ber. Verh. Sächs. Akad. Wiss. Leipzig Math.-Phys. Kl.* **69** (1917), 306–318.
2. W. Blaschke, Affine Geometrie IX: Verschiedene Bemerkungen und Aufgaben, *Ber. Verh. Sächs. Akad. Wiss. Leipzig Math.-Phys. Kl.* **69** (1917), 412–420.

3. W. Blaschke, Affine Geometrie XIV: Eine Minimumaufgabe für Legendres Trägheitsellipsoid, *Ber. Verh. Sächs. Akad. Wiss. Leipzig Math.-Phys. Kl.* **70** (1918), 72–75.
4. J. Bourgain and V. Milman, New volume ratio properties for convex symmetric bodies in $\mathbb{R}^n$, *Invent. Math.* **88** (1987), 319–340.
5. N. S. Brannen, Volumes of projection bodies, *Mathematika* **43** (1996), 255–264.
6. Yu. D. Burago and V. A. Zalgaller, *Geometric Inequalities*, Springer-Verlag, Berlin Heidelberg, 1988.
7. S. Campi, A. Colesanti and P. Gronchi, A note on Sylvester's problem for random polytopes in a convex body, *Rend. Ist. Mat. Univ. Trieste* **31** (1999), 79–94.
8. S. Campi and P. Gronchi, The L^p-Busemann-Petty centroid inequality, *Adv.Math.* **167** (2002), 128–141.
9. S. Campi and P. Gronchi, On the reverse L^p-Busemann-Petty centroid inequality, *Mathematika* **49** (2002), 1–11.
10. S. Campi and P. Gronchi, On volume product inequalities for convex sets, *preprint.*
11. R. J. Gardner, *Geometric Tomography*, Cambridge University Press, Cambridge, 1995.
12. P. Goodey and W. Weil, Zonoids and Generalisations, in *Handbook of Convex Geometry* (eds. P. M. Gruber and J. M. Wills), North-Holland, Amsterdam, 1993, 1297–1326.
13. Y. Gordon, M. Meyer and S. Reisner, Zonoids with minimal volume-product - A new proof, *Proc. AMS* **104** (1988), 273–276.
14. F. John, Polar correspondance with respect to convex regions, *Duke Math. J.* **3** (1937), 355–369.
15. F. John, Extremum problems with inequalities as subsidiary conditions, in: Courant Anniversary Volume (Interscience, New York), 1948, pp. 187–204.
16. E. Lutwak, On a conjectured projection inequality of Petty, *Contemp. Math.* **113** (1990), 171–182.
17. E. Lutwak, Selected affine isoperimetric inequalities, in *Handbook of Convex Geometry* (eds. P. M. Gruber and J. M. Wills), North-Holland, Amsterdam, 1993, 151–176.
18. E. Lutwak, D. Yang and G. Zhang, L_p affine isoperimetric inequalities, *J. Differential Geom.* **56** (2000), 111–132.
19. E. Lutwak and G. Zhang, Blaschke-Santaló inequalities, *J. Differential Geom.* **47** (1997), 1–16.
20. K. Mahler, Ein Übertragungsprinzip für konvexe Körper, Časopis Pěst. Mat. Fys. **68** (1939), 93–102.
21. K. Mahler, Ein Minimalproblem für konvexe Polygone, *Mathematica (Zutphen) B* **7** (1939), 118–127.
22. M. Meyer, Convex bodies with minimal volume product in $\mathbb{R}^2$, *Monatsh. Math.* **112** (1991), 297–301.
23. V. D. Milman and A. Pajor, Isotropic position and inertia ellipsoids and zonoids of the unit ball of a normed n-dimensional space, in *Geometric*

Aspects of Functional Analysis (eds. J. Lindenstrauss and V. D. Milman), Lecture Notes in Mathematics **1376**, Springer, Heidelberg, 1989, 64–104.
24. C. M. Petty, Centroid surfaces, *Pacific J. Math.* **11** (1961), 1535–1547.
25. C. M. Petty, Isoperimetric problems, in *Proc. Conf. Convexity and Combinatorial Geomerty, Univ. Oklahoma*, (1971), 26–41.
26. C. M. Petty, Affine isoperimetric problems, *Ann. N. Y. Acad. Sc.* **440** (1985), 113–127.
27. S. Reisner, Random polytopes and the volume product of symmetric convex bodies, *Math. Scand.* **57** (1985), 386–392.
28. S. Reisner, Zonoids with minimal volume product, *Math. Z.* **192** (1986), 339–346.
29. C. A. Rogers and G. C. Shephard, Some extremal problems for convex bodies, *Mathematika* **5** (1958), 93–102.
30. J. Saint-Raymond, Sur le volume des corps convexes symétriques, in *Séminaire Choquet - Initiation à l'Analyse 1980/81 Exp. No. 11*, Université Pierre et Marie Curie, Paris, 1981, 1–25.
31. L. A. Santaló, Un invariante afin para los cuerpos convexos del espacio de n dimensiones, *Portugal. Math.* **8** (1949), 155–161.
32. R. Schneider, *Convex bodies: the Brunn-Minkowski theory*, Cambridge University Press, Cambridge, 1993.
33. G. C. Shephard, Shadow systems of convex bodies, *Israel J. Math.* **2** (1964), 229–36.
34. G. Zhang, Restricted chord projection and affine inequalities, *Geom. Dedicata* **39** (1991), 213–222.

INTEGRAL GEOMETRY AND ALESKER'S THEORY OF VALUATIONS

JOSEPH H.G. FU *

Department of Mathematics
University of Georgia
Athens GA 30602, USA
E-mail: fu@math.uga.edu

Let G be a closed subgroup of $SO(n)$ acting transitively on the unit sphere S^{n-1}. Then there exists an array of "kinematic formulas" for G, analogous to Blaschke's Principal Kinematic Formula for the $SO(n)$, where the full group of oriented euclidean motions is replaced by those whose linear part lies in G. This array is part of an intriguing algebraic structure based on the Irreducibility Theorem and the multiplicative structure on valuations, due to S. Alesker.

1. Introduction

The famous Principal Kinematic Formula of Blaschke states that if $A, B \subset \mathbb{R}^n$ are convex bodies and $\overline{SO(n)} := SO(n) \ltimes \mathbb{R}^n$ is the oriented euclidean group and $\Phi_0, \dots \Phi_n$ are the intrinsic volumes, indexed by degree, then

$$\int_{\overline{SO(n)}} \Phi_k(A \cap \bar{g}B) \, d\bar{g} = \sum_{i+j=n+k} c_{ij}^k \Phi_i(A) \Phi_j(B), \tag{1}$$

for some universal constants c_{ij}^k.

Historically there have basically two different proofs of the existence of these formulas, each offering a different view of the situation. One proof begins with the observation that the intrinsic volumes of a convex set A may be expressed as integrals of certain differential forms in the sphere bundle $\mathbb{R}^n \times S^{n-1}$, invariant under the induced action of $\overline{SO(n)}$, over the "normal cycle" $N(A)$, the manifold of unit normals to A. The proof then proceeds by the classic integral geometric device of constructing a kind of incidence

*Work partially supported by the National Science Foundation. The author also wishes to thank Charles University in Prague for its gracious hospitality as this paper was written.

correspondence for the pair A, B, living in the product $\mathbb{R}^n \times S^{n-1} \times \overline{SO(n)}$ of the sphere bundle with the group, which when sliced by the projection to $\overline{SO(n)}$ gives the normal cycles $N(A \cap \bar{g}B), \overline{G} \in \overline{SO(n)}$[1]. This approach permits one to generalize the formula (1) by enlarging the class of sets admitting normal cycles (although it should be emphasized that this larger class remains poorly understood[2]). A different kind of generalization is also possible by considering a smaller subgroup $G \subset SO(n)$. This is possible provided the action of G is transitive on the sphere S^{n-1}, since this is precisely what is necessary to ensure that the space of invariant forms is finite dimensional.

The other approach proceeds via Hadwiger's classic theorem on $\overline{SO(n)}$-invariant convex valuations. Let $\mathcal{K}^n$ denote the space of all compact convex subsets of $\mathbb{R}^n$, endowed with the Hausdorff metric. Then $\mathcal{K}^n$ is locally compact (the "Blaschke Selection Theorem"). Put $\mathbf{Val} = \mathbf{Val}(\mathbb{R}^n)$ for the vector space of all continuous real-valued translation-invariant convex valuations on $\mathbb{R}^n$, i.e. the space of all continuous functions $\varphi : \mathcal{K}^n \to \mathbb{R}$ such that

- $\varphi(\emptyset) = 0$,
- if $x \in \mathbb{R}^n$ and $A \in \mathcal{K}^n$ then $\varphi(A + x) = \varphi(A)$, and
- if $A, B, A \cup B \in \mathcal{K}^n$ then $\varphi(A \cup B) = \varphi(A) + \varphi(B) - \varphi(A \cap B)$.

We will refer to such functionals φ simply as valuations. The key fact is then

Theorem 1.1. *(Hadwiger) The space* $\mathbf{Val}^{SO(n)}$ *of* $\overline{SO(n)}$*-invariant, continuous convex valuations is finite-dimensional, and the intrinsic volumes* $\Phi_0, \ldots, \Phi_n$ *constitute a basis.*

The kinematic formulas (1) follow immediately, since the left-hand side defines an invariant valuation in A, B separately.

In each of these approaches, once the validity of some such formula is established the precise values of the constants c_{ij}^k are then computed from convenient choices of the sets A, B. The connection between the two approaches arises from the fact that, as an operator with values lying in the space of integral currents of the sphere bundle $\mathbb{R}^n \times S^{n-1}$, the normal cycle N is also a valuation in the sense that $\varphi = N$ satisfies the last condition listed above.

Recently the work of S. Alesker has revealed that the second approach (based on Hadwiger's theorem) is also available for other groups $G \subset SO(n)$ as above. It also opens the door to the study of a rich algebraic structure, whose general characteristics we describe in the rest of this article. This area is still in its infancy, with many seemingly accessible open problems.

2. Some naïve algebra

We begin with a few completely formal observations. The formulas (1) may be regarded as giving a comultiplication on $\mathbf{Val}^{SO(n)}$, i.e. a linear map $k = k_{SO(n)} : \mathbf{Val}^{SO(n)} \to \mathbf{Val}^{SO(n)} \otimes \mathbf{Val}^{SO(n)}$ defined by

$$\mathbf{Val}^{SO(n)}(\Phi_k) := \sum_{i+j=n+k} c_{ij}^k \Phi_i \otimes \Phi_j, \tag{2}$$

where the structure constants c_{ij}^k are taken from (1).

In fact this operation is cocommutative, i.e. the image of k in $\mathbf{Val}^{SO(n)} \otimes \mathbf{Val}^{SO(n)}$ is symmetric. This is obvious since $A \cap \bar{g}B$ is congruent to $\bar{g}^{-1}A \cap B$, and the map $\bar{g} \mapsto \bar{g}^{-1}$ preserves the Haar measure of $\overline{SO(n)}$. Furthermore the comultiplication is coassociative, in the sense that the diagram

$$\begin{array}{ccc} \mathbf{Val}^{SO(n)} & \xrightarrow{k} & \mathbf{Val}^{SO(n)} \otimes \mathbf{Val}^{SO(n)} \\ {\scriptstyle k}\downarrow & & \downarrow{\scriptstyle k\otimes id} \\ \mathbf{Val}^{SO(n)} \otimes \mathbf{Val}^{SO(n)} & \xrightarrow{id\otimes k} & \mathbf{Val}^{SO(n)} \otimes \mathbf{Val}^{SO(n)} \otimes \mathbf{Val}^{SO(n)} \end{array} \tag{3}$$

commutes. This follows at once from Fubini's theorem:

$$\begin{aligned} \int\int_{\overline{SO(n)}^2} \Phi_k(A \cap \bar{g}B \cap \bar{h}C)\, d\bar{g}\, d\bar{h} &= \sum c_{ij}^k \left(\int_{\overline{SO(n)}} \Phi_i(A \cap \bar{g}B)\, d\bar{g} \right) \Phi_j(C) \\ &= \sum c_{ij}^k \Phi_i(A) \left(\int_{\overline{SO(n)}} \Phi_j(B \cap \bar{h}C)\, d\bar{h} \right). \end{aligned}$$

This is a convenient language in which to state a striking fact due to Nijenhuis [3]:

Theorem 2.1. *There are constants γ_i and a choice of normalization for the Haar measure $d\bar{g}$ such that, putting $\Psi_i := \gamma_i \Phi_i$, $i = 0, \dots, n$,*

$$k_{SO(n)}(\Psi_k) = \sum_{i+j=n+k} \Psi_i \otimes \Psi_j. \tag{4}$$

That is, with respect to the basis Ψ_i, all the structure constants for the kinematic comultiplication are equal to unity.

3. Consequences of the Irreducibility Theorem

We recall a fundamental theorem due to P. McMullen. Say that $\varphi \in \mathbf{Val}(R^n)$ has **degree** i if $\varphi(rA) = r^i\varphi(A)$ whenever $r \geq 0$. Put $\mathbf{Val}_i \subset \mathbf{Val}$ for the space of all valuations of degree i.

Theorem 3.1. *(P. McMullen [4])*

$$\mathbf{Val}(\mathbb{R}^n) = \bigoplus_{i=0}^{n} \mathbf{Val}_i(\mathbb{R}^n). \tag{5}$$

Furthermore $\mathbf{Val}_n(\mathbb{R}^n)$ *is one-dimensional, and is spanned by the volume functional* $\mathbf{vol}^n$. *Similarly* $\mathbf{Val}_0(\mathbb{R}^n)$ *is also one-dimensional, and is spanned by the Euler characteristic* χ.

Observe that χ reduces to the constant function 1 on $\mathcal{K}^n$; in the literature it is referred to as χ because its natural extension to larger classes of sets coincides with the Euler characteristic.

We say that $\varphi \in \mathbf{Val}$ is **even** (resp. **odd**) if $\varphi(-A) = (-1)^\epsilon\varphi(A)$, $\epsilon = 0$ (resp. $\epsilon = 1$) for all $A \in \mathcal{K}$. Clearly

$$\mathbf{Val}_i = \mathbf{Val}_i^0 \oplus \mathbf{Val}_i^1$$

where $\mathbf{Val}^\epsilon$ denotes the space of valuations of parity ϵ. We observe that $GL(n)$ acts naturally on $\mathbf{Val}$ by

$$(g\varphi)(A) := \varphi(g^{-1}A).$$

Furthermore $\mathbf{Val}$ carries the structure of a Banach space under the norm

$$\| \varphi \| := \sup\{|\varphi(A)| : A \subset B(0,1)\}.$$

It is clear that $\mathbf{Val}$ decomposes as the direct sum of the $\mathbf{Val}_i^\epsilon$, and that each of these is a continuous representation of $GL(n)$. Alesker's Irreducibility Theorem states that this gives a complete decomposition of the full representation $\mathbf{Val}$:

Theorem 3.2. *(Alesker [5]) Each representation* $\mathbf{Val}_i^\epsilon$, $i = 0, \ldots, n$, $\epsilon = 0, 1$, *is irreducible, i.e. it admits no closed proper subrepresentations.*

For our purposes, this theorem has the following main consequences. We begin by defining, for $A \in \mathcal{K}$, the valuation μ_A by

$$\mu_A(B) := \mathbf{vol}^n(B + (-A)).$$

Note also that

$$\mu_A(B) = \int_{\mathbb{R}^n} \chi(B \cap (A + x))\, dx. \tag{6}$$

Using this result, Alesker proved the following conjecture of McMullen:

Corollary 3.1. *The valuations $\mu_A, A \in \mathcal{K}$, span a dense subspace of* **Val**.

Proof. By dilating, renormalizing and decomposing into odd and even parts, it is clear that any closed subrepresentation of **Val** is a direct sum of closed subrepresentations of the $\mathbf{Val}_i^\epsilon$. By the Irreducibility Theorem, these are either zero or the $\mathbf{Val}_i^\epsilon$ themselves. But it is clear that the closure of the span of the valuations μ_A has nontrivial components in every degree and parity. □

Corollary 3.2. *Given a smooth differential form α of degree $n-1$ on the sphere bundle $\mathbb{R}^n \times S^{n-1}$, invariant under translations, define $\nu_\alpha \in$* **Val** *by*

$$\nu_\alpha(A) := \int_{N(A)} \alpha.$$

Then the span of the valuations ν_α is dense in $\bigoplus_{i<n} \mathbf{Val}_i$.

Proof. Again it is easy to see that the span of such valuations includes nonzero subspaces of each of the $\mathbf{Val}_i^\epsilon$. Thus the Irreducibility Theorem implies that the span must be dense. □

From this point on we let G denote a closed subgroup of $SO(n)$ that acts transitively on S^{n-1}. The results above have profound consequences for the space $\mathbf{Val}^G$ of G-invariant valuations. We put $\overline{G} := G \ltimes \mathbb{R}^n$ for the semidirect product with the translation group.

Corollary 3.3. *For such G, the space $\mathbf{Val}^G$ is finite-dimensional, and is equal to the direct sum of $\mathbf{Val}_n$ with space of all valuations $\nu_{\bar\alpha}$ as $\bar\alpha$ ranges over the space of all $\overline{G}$-invariant differential forms of degree $n-1$ on the sphere bundle of $\mathbb{R}^n$.*

Proof. Note first that the space W of all valuations of the form $\nu_{\bar\alpha}$, where $\bar\alpha$ is a differential form on $\mathbb{R}^n \times S^{n-1}$ invariant under $\overline{G}$, is dense in $V := \bigoplus_{i<n} \mathbf{Val}^G$. For if $\varphi \in \mathbf{Val}^G$, $\deg \varphi < n$, then given $\epsilon > 0$ there is a translation-invariant form α such that $\| \varphi - \nu_\alpha \| < \epsilon$. Averaging over G, since φ is G-invariant we find that $\| \varphi - \nu_{\bar\alpha} \| < \epsilon$ as well, where

$$\bar\alpha := dg(G)^{-1} \int_G g^* \alpha \, dg.$$

But W is finite-dimensional, so in fact $W = V$. □

Corollary 3.4. *For G as above, put*

$$\mu_A^G := \int_G g\mu_A \, dg.$$

Then $\mathbf{Val}^G$ *is spanned by the valuations* μ_A^G.

Proof. Since $\mathbf{Val}^G$ is finite-dimensional by Corollary 3.3, the present statement may be proved by a similar argument. □

Now we may prove the following, the first part of which is a restatement of a theorem of Alesker [5].

Theorem 3.3. *Let $G \subset SO(n)$ be a closed subgroup acting transitively on $SO(n)$. Then there is a cocommutative, coassociative coproduct k_G : $\mathbf{Val}^G \to \mathbf{Val}^G \otimes \mathbf{Val}^G$, of degree n, such that for $\varphi \in \mathbf{Val}^G$*

$$\int_{\overline{G}} \varphi(A \cap \bar{g}B) \, d\bar{g} = k_G(\varphi)(A, B). \tag{7}$$

Furthermore, if we identify $\mathbf{Val}^G \otimes \mathbf{Val}^G$ in the natural way with $\mathrm{Hom}_{\mathbb{R}}((\mathbf{Val}^G)^*, \mathbf{Val}^G)$ then $k_G(\chi) : (\mathbf{Val}^G)^* \to \mathbf{Val}^G$ is an isomorphism of graded vector spaces.

Proof. The existence of such a formula for the integral in (7) follows from the finite-dimensionality of $\mathbf{Val}^G$, just as in the proof of (1) via Hadwiger's theorem. The algebraic properties of the resulting comultiplication k_G may be proved as in Section 2. That k_G has degree n follows from a simple scaling argument.

To show that $k_G(\chi)$ is an isomorphism it is enough to show that it is surjective, and by Corollary 3.4 it is enough to show that every valuation μ_A^G lies in the image. In fact, putting $e_A \in (\mathbf{Val}^G)^*$ to be the evaluation function $e_A(\varphi) := \varphi(A)$, we have for every $B \in \mathcal{K}$

$$\begin{aligned} k_G(\chi)(e_A)(B) &= \int_{\overline{G}} \chi(B \cap \bar{g}A) \, d\bar{g} \\ &= \mu_A^G(B) \end{aligned}$$

by (6).

That the map is graded follows at once from the fact that the degree of k_G is n. □

4. Alesker multiplication

The other major tool that Alesker's approach offers is the definition of a commutative graded product on a dense subspace of **Val**.

Definition 4.1. (Alesker [6]) Given $A, B \in \mathcal{K}$, the product of the valuations μ_A, μ_B is defined by

$$(\mu_A \cdot \mu_B)(C) := \mathbf{vol}^{2n}\left(\Delta_C + (A \times \{0\}) + (\{0\} \times B)\right)$$

where $\Delta_C := \{(c,c) : c \in C\} \subset \mathbb{R}^{2n}$ is the diagonal embedding of C. If $G \subset SO(n)$ is a compact subgroup then this product may be extended by continuity to

$$(\mu_A^G \cdot \mu_B^G)(C) := \int\int_{G\times G} \mathbf{vol}^{2n}\left(\Delta_C + (gA \times \{0\}) + (\{0\} \times hB)\right) dg\, dh.$$

Using Fubini's theorem it is easy to rewrite the product in the asymmetric form

$$(\mu_A \cdot \mu_B)(C) = \int_{\mathbb{R}^n} \mu_B(C \cap (A + x))\, dx.$$

Extending by continuity in the second factor, we may therefore write also for general $\varphi \in \mathbf{Val}$

$$(\varphi \cdot \mu_A)(C) = (\mu_A \cdot \varphi)(C) := \int_{\mathbb{R}^n} \varphi(C \cap (A + x))\, dx. \tag{8}$$

and similarly

$$\begin{aligned} (\mu_A^G \cdot \varphi)(C) &= \int_G \int_{\mathbb{R}^n} \varphi(C \cap (gA + x))\, dx \\ &= \int_{\overline{G}} \varphi(C \cap \bar{g}A)\, d\bar{g}. \end{aligned} \tag{9}$$

Clearly this product is commutative and associative, and the relations (6), (8) and (9) show that the valuation χ acts as the multiplicative identity. It is not hard to show also that the product respects the grading of **Val**, in the sense that $\deg(\varphi \cdot \psi) = \deg \varphi + \deg \psi$. The Irreducibility Theorem also implies that for every $\varphi \in \mathbf{Val}$, there exists $A \in \mathcal{K}$ such that $\mu_A \cdot \varphi \neq 0$. Restricting to $\mathbf{Val}^G$, this implies that $\mathbf{Val}^G$ is a "Frobenius algebra":

Theorem 4.1. *The "Poincaré duality" pairing* PD : $\mathbf{Val}^G \times \mathbf{Val}^G \to \mathbf{Val}_n^G \simeq \mathbb{R}$ *given by*

$$(\varphi, \psi) \mapsto (\varphi \cdot \psi)_n$$

(degree n part) is perfect.

Note that the Poincaré duality pairing may be regarded as an isomorphism PD : $\mathbf{Val}^G \to (\mathbf{Val}^G)^*$ of graded vector spaces. In fact it is clear that it is even an isomorphism of $\mathbf{Val}^G$-modules, where the $\mathbf{Val}^G$-module structure of $(\mathbf{Val}^G)^*$ is given by

$$(\varphi \cdot \theta^*)(\psi) := \theta^*(\varphi \cdot \psi)$$

for $\varphi, \psi \in \mathbf{Val}^G$ and $\theta^* \in (\mathbf{Val}^G)^*$.

The relation (9) suggests that the product structure on $\mathbf{Val}^G$ is closely related to the kinematic comultiplication k_G. The precise relation may be encoded as follows.

Theorem 4.2. *If $\varphi, \psi \in \mathbf{Val}^G$ then*

$$k_G(\varphi \cdot \psi) = (\varphi \otimes 1) \cdot k_G(\psi) = (1 \otimes \varphi) \cdot k_G(\psi).$$

Proof. By Corollary 3.4, it is enough to prove this for $\varphi = \mu_A^G$, $A \in \mathcal{K}$. Then for $B, C \in \mathcal{K}$

$$\begin{aligned} k_G(\mu_A^G \cdot \psi)(B, C) &= \int_{\overline{G}} (\mu_A^G \cdot \psi)(B \cap \bar{g}C)\, d\bar{g} \\ &= \int_{\overline{G}} \int_{\overline{G}} \psi((B \cap \bar{g}C) \cap \bar{h}A)\, d\bar{g}\, d\bar{h} \\ &= \int_{\overline{G}} k_G(\psi)(B \cap \bar{h}A, C)\, d\bar{h} \\ &= (\mu_A^G \otimes 1) \cdot k_G(\psi)(B, C). \end{aligned}$$

Hence $k_G(\mu_A^G \cdot \psi) = (\mu_A^G \otimes 1) \cdot k_G(\psi)$, and similarly $k_G(\mu_A^G \cdot \psi) = (1 \otimes \mu_A^G) \cdot k_G(\psi)$. □

Corollary 4.1. *$k_G(1) = k_G(\chi) \in \mathrm{Hom}_{\mathbb{R}}((\mathbf{Val}^G)^*, \mathbf{Val}^G)$ is in fact an isomorphism of graded $\mathbf{Val}^G$-modules.*

Proof. Thinking of $k_G(1)$ as homomorphism $(\mathbf{Val}^G)^* \to \mathbf{Val}^G$, and putting m_φ to be multiplication by $\varphi \in \mathbf{Val}^G$, we compute

$$\begin{aligned} k_G(1) \circ m_\varphi &= k_G(1) \cdot (1 \otimes \varphi) \\ &= (\varphi \otimes 1) \cdot k_G(1) \\ &= m_\varphi \circ k_G(1), \end{aligned}$$

by Theorem 4.2, as claimed. □

Corollary 4.2. *Up to a constant scaling factor, $k_G(1) = \mathrm{PD}^{-1}$.*

Proof. Clearly any $\mathbf{Val}^G$-map $\mathbf{Val}^G \to (\mathbf{Val}^G)^*$ is determined by the value of the multiplicative identity $\chi = 1$, and $k_G(1)^{-1}(1)$ and PD(1) are both nonzero elements of the one-dimensional space $(\mathbf{Val}_n^G)^*$. Hence they must be multiples of each other. □

5. Applications to the integral geometry of the orthogonal and unitary groups

The following elaboration of the Hadwiger theorem is due to Alesker.

Theorem 5.1. *The algebra* $\mathbf{Val}^{SO(n)}$ *is isomorphic to* $\mathbb{R}[t]/(t^{n+1})$.

Proof. Here the generator $t = \Phi_1$ is the mean breadth, i.e.

$$t(A) := \int_{G(n,1)} \mathrm{length}(\pi_l(A))\, dl \tag{10}$$

where $G(n,1)$ is the projective space of all lines l through the origin and π_l denotes the orthogonal projection onto l. This valuation may be viewed as a limit of valuations of the form $\mu_B^{SO(n)}$ as follows. Put $\bar{G}(n,j)$ for the space of all affine subspaces of dimension j in $\mathbb{R}^{n-1}$. Then clearly

$$\begin{aligned} t(A) &= \int_{\bar{G}(n,n-1)} \chi(A \cap \bar{P})\, d\bar{P} \\ &= \lim_{r\to\infty} r^{1-n} \mu_{D_r}^{SO(n)}(A), \end{aligned}$$

where D_r is an $(n-1)$-dimensional disk of radius r. Using the expression (9) for the Alesker product it follows that

$$\begin{aligned} t^k(A) &= \int_{\bar{G}(n,n-1)^k} \chi(A \cap \bar{P}_1 \cap \cdots \cap \bar{P}_k)\, d\bar{P}_1 \ldots d\bar{P}_k \\ &= \int_{\bar{G}(n,n-k)} \chi(A \cap \bar{Q})\, d\bar{Q} \\ &= \Phi_k(A). \end{aligned}$$

□

Clearly the Poincaré duality pairing identifies t^i with the dual basis element $(t^{n-i})^*$, i.e. $\mathrm{PD} = \sum_{i=0}^n (t^i)^* \otimes (t^{n-i})^*$ as an element of $(\mathbf{Val}^G)^* \otimes (\mathbf{Val}^G)^*$. Thus Corollary 4.2 yields

$$k_{SO(n)}(\chi) = \sum_{i=0}^{n} t^i \otimes t^{n-i},$$

and Theorem 4.2 then implies that

$$k_{SO(n)}(t^k) = \sum_{i+j=n+k} t^i \otimes t^j, \ k = 0, \ldots, n,$$

thus proving Nijenhuis's Theorem 2.1.

Thus from this perspective the case $G = SO(n)$ becomes the trivial ground case of a theory yet to be worked out for the other compact groups G that act transitively on the sphere. The first nontrivial case is $G = U(n)$, acting on the sphere S^{2n-1} of the euclidean space $\mathbb{C}^n \simeq \mathbb{R}^{2n}$. In this case we have the following theorem:

Theorem 5.2. [7] *The graded algebra* $\mathbf{Val}^{U(n)}$ *is isomorphic to* $\mathbb{R}[s,t]/(f_{n+1}, f_{n+2})$, *where* $\deg s = 2$, $\deg t = 1$ *and the* f_i *are the components of weighted degree* i *in the expansion*

$$\log(1 + s + t) = \sum_{i=1}^{\infty} f_i.$$

In principle, this fact determines the Poincar'e duality pairing and therefore also the array of kinematic formulas for the unitary group. In view of the extremely simple structure of the algebra $\mathbf{Val}^{SO(n)}$, it was easy at this stage to write down these formulas in closed form in that case. So far we have not been able to find any such general closed forms in the unitary case.

There are many such open questions about unitary-invariant valuations, suggested by more or less trivial facts from the $SO(n)$ case. This area deserves further study.

References

1. J.H.G. Fu, Kinematic formulas in integral geometry. *Indiana U. Math. J.*, **39** (1990).
2. J.H.G. Fu, Curvature measures of subanalytic sets. *Amer. J. Math.* **116** (1994).
3. A. Nijenhuis, On Chern's kinematic formula in integral geometry. *J. Differential Geom.*, **9** (1974).
4. P. McMullen, Valuations and Euler-type relations on certain classes of convex polytopes. *Proc. London Math. Soc.*, **35** (1977).
5. S. Alesker, Deescription of translation invariant valuations on convex sets with solution of P. McMullen's conjecture. *Geom. Funct. Anal.* **11** (2001).

6. S. Alesker, The multiplicative structure on polynomial continuous valuations. *Geom. Funct. Anal.* **14** (2004).
7. J.H.G. Fu, Structure of the unitary valuation algebra. arXiv:math.DG/0410575

AREA AND PERIMETER BISECTORS OF PLANAR CONVEX SETS

PAUL GOODEY

University of Oklahoma,
Norman,
Oklahoma 73019, USA

This paper collects some results concerning lengths of area and perimeter bisecting chords of planar convex sets. Particular attention is paid to results of an integral geometric nature as well as some early work by Auerbach, who used Fourier analysis techniques. These results will be seen to provide some answers to more recent questions.

1. Introduction

Our considerations are motivated by problem **A26** of [5]. This question asks for information about the lengths of those chords of a planar convex set K which bisect the area $A(K)$ of K. In particular, they mention a problem of Santaló which asks whether there is always such a chord of length at most $2\sqrt{A(K)/\pi}$. They also ask whether other bounds on the length of these chords can be found in terms of, for example, the mean width, perimeter length or diameter of K. In this paper we will provide a brief survey of some of the known results and show, in particular, that they provide a negative answer to Santaló's question. However, it will be apparent that there is always an area bisecting chord of length at least $2\sqrt{A(K)/\pi}$. Our primary reference will be the work of Auerbach [1] which provides an excellent example of the early use of Fourier analysis to produce an intriguing class of convex sets. Following Auerbach, we will also consider chords which bisect the perimeter of K.

2. Notation and some early results

As already indicated, we will denote by $A(K)$ the area of a planar convex set K. Its perimeter length will be denoted by $L(K)$. The largest radius of a circle contained in K is the inradius and will be written $r(K)$. Similarly, the radius of the smallest circle containing K is the circumradius, written $R(K)$. For a given direction $\theta \in [0, 2\pi]$, the width, $w(K, \theta)$ of K in direction θ is the distance between the supporting lines perpendicular to θ. The minimum width of K is denoted $b(K)$ and the maximum width is the diameter, denoted $D(K)$. It is easy to see that, for each $\theta \in [0, 2\pi]$, there will be precisely one chord parallel to θ which bisects the area of K. The length of this chord will be denoted by $A(K, \theta)$. Similarly, the length of the perimeter bisecting chord in this direction will be written $L(K, \theta)$.

In [12], Pál observes that if a convex set has minimum width b then its area is at least $b^2/\sqrt{3}$. In fact, he also shows that the equilateral triangle provides the only case of minimum area. If we denote by $\theta_0 \in [0, 2\pi]$ a direction in which the minimum width is achieved, then $A(K, \theta_0) \leq b$ and so

$$A(K) \geq \frac{b^2}{\sqrt{3}} \geq \frac{A(K, \theta_0)^2}{\sqrt{3}}.$$

So there is always an area bisecting chord of length at most $3^{1/4}\sqrt{A(K)}$. Of course, the same can be said of perimeter bisecting chords.

Radziszewski [13] and [14] found lower bounds for the maximum lengths of area and perimeter bisecting chords. He showed that there is always an area bisecting chord of length at least $3D(K)/4$ and a perimeter bisecting chord of length at least $\lambda_0 D(K)$ where $\lambda_0 \approx 0.829$. Eggleston [6] obtained similar bounds in terms of the circumradius. In order to do this, he improved Radziszewski's area bisecting result. He put

$$f(x, y) = \begin{cases} \left(\frac{9}{16}x^2 + \frac{1}{4}y^2\right)^{1/2} & \text{if } y^2 \leq \frac{3}{8}x^2 \\ \sqrt{\frac{21}{32}}x & \text{if } y^2 > \frac{3}{8}x^2 \end{cases}$$

and showed that there is always an area bisecting chord of length at least $f(D(K), b(K))$. In case $b(K)^2 \leq 3D(K)^2/8$, this result is best possible. He was able to use these results to show that there is always an area bisecting chord of length at least $3R(K)/2$. He also used Radziszewski's results to show that there is always a perimeter bisecting chord of length at least $3R(K)/2$.

3. Auerbach's results

It is an elementary fact that, in the case of a centrally symmetric set K, every chord through the centre is simultaneously area and perimeter bisecting. Auerbach [1] constructed examples of sets K, which are not centrally symmetric, for which the area and perimeter bisecting chords are the same and have constant length. He points out that this phenomenon had already been observed by Zindler [17]. These examples provide a negative answer to Santaló's question and so we shall give an outline of his approach.

Auerbach investigates convex sets K which have the property that every chord of some given length, σ say, determines an arc of fixed length. He first shows that, if there are such examples, then these same chords determine sectors of fixed area - hence his interest in floating bodies and Ulam's problem. He considers bodies with smooth boundaries and investigates the angle $\alpha(\theta)$ between the area (and perimeter) bisecting chord in direction θ and the tangent at the point from which the chord emanates. He shows that:

- $0 < \alpha(\theta) < \pi$ for all $\theta \in [0, 2\pi]$;
- $\alpha(\theta) + \alpha(\theta + \pi) = \pi$ for all $\theta \in [0, 2\pi]$;
- α satisfies the Lipschitz condition $|\alpha(\theta_2) - \alpha(\theta_1)| \leq |\theta_2 - \theta_1|$;
- the Fourier series expansion of $\cot \alpha(\theta)$ has no terms of degree less than three.

The Fourier series expansion of $\cot \alpha(\theta)$ can therefore be written in the form

$$\cot \alpha(\theta) \sim \sum_{k=1}^{\infty} (a_{2k+1} \cos(2k+1)\theta + b_{2k+1} \sin(2k+1)\theta). \tag{1}$$

Conversely, if the a_{2k+1} and b_{2k+1} are sufficiently close to zero, then the angle θ defined by (1) will have the properties itemized above. He then fixes a positive number σ and considers the curve with parametric equations

$$x(\theta) = -\frac{\sigma}{2} \cos\theta + \frac{\sigma}{2} \sum_{k=1}^{\infty} \left(-\frac{b_{2k+1} + b_{2k-1}}{4k} \cos 2k\theta + \frac{a_{2k+1} + a_{2k-1}}{4k} \sin 2k\theta \right)$$

$$y(\theta) = -\frac{\sigma}{2} \sin\theta + \frac{\sigma}{2} \sum_{k=1}^{\infty} \left(\frac{a_{2k+1} - a_{2k-1}}{4k} \cos 2k\theta + \frac{b_{2k+1} - b_{2k-1}}{4k} \sin 2k\theta \right).$$

He shows that this is, indeed, a closed convex curve C. Furthermore, for each $\theta \in [0, 2\pi]$, the chord emanating from the point $(x(\theta), y(\theta))$ and making angle $\alpha(\theta)$ with the tangent at $(x(\theta), y(\theta))$ is of length σ and bisects both the area and perimeter of the convex set K with boundary C. The following diagram shows an example of such a convex set and some of its area and

perimeter bisecting chords

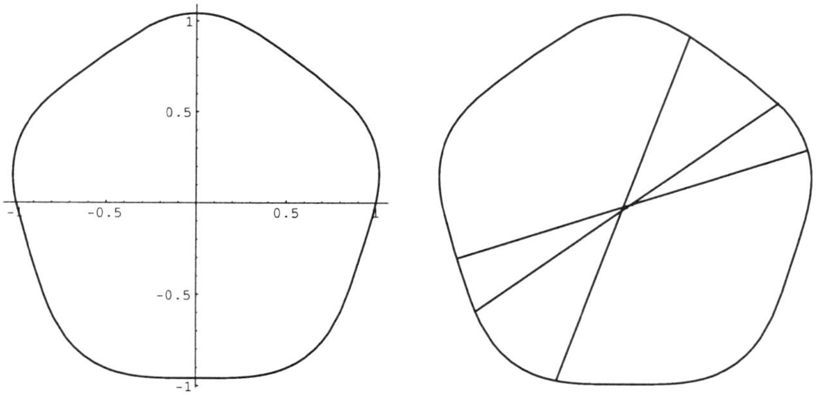

Using the formulas of Green and Parseval, Auerbach shows that

$$A(K) = \pi \frac{\sigma^2}{8} \left(2 - \sum_{k=1}^{\infty} \frac{1}{k(k+1)} \left(a_{2k+1}^2 + b_{2k+1}^2 \right) \right).$$

It is clear that this provides a negative answer to Santaló's question, since $\sigma > 2\sqrt{A(K)/\pi}$.

4. Some integral geometry results

Perhaps the initial use of integral geometry for these problems was motivated by a conjecture of Herda [9]. He conjectured that the minimum length of a perimeter bisecting chord is at most $L(K)/\pi$ with equality holding only for circular discs. This conjecture was confirmed by Ault [2], Chakerian [3], Goodey [7], Witsenhausen [16] and others, see [10].

Both Chakerian and Goodey established the somewhat stronger result

$$\frac{1}{2\pi} \int_0^{2\pi} L(K, \theta)\, d\theta \leq \frac{1}{\pi} L(K),$$

with equality holding only for circular discs. This result was then improved and extended to area bisectors by Lutwak [11]. He considered L^2 averages and showed that

$$\left(\frac{1}{2\pi} \int_0^{2\pi} L^2(K, \theta)\, d\theta \right)^{1/2} \leq \frac{L(K)}{\pi}$$

and

$$\left(\frac{1}{2\pi}\int_0^{2\pi} A^2(K,\theta)\,d\theta\right)^{1/2} \le \frac{L(K)}{\pi}.$$

He also showed that equality occurs in either of these only in the case that K is a circular disc. As he indicated, these results show that the minimum length of an area bisecting chord is at most $L(K)/\pi$, with equality only for discs.

Lutwak's results were subsequently extended by Chakerian and Goodey [4] who considered *simple* families of chords. A family $\mathcal{F}$ of chords of a planar convex set K is said to be simple if, for each direction θ, there is just one member $f(\theta) \in \mathcal{F}$ lying in that direction and if the end-points of the chords $f(\theta)$ each traverse the boundary of K continuously in the counter clockwise direction as θ varies from 0 to 2π. They showed that, if $\mathcal{F}$ is a simple family of chords and if the length of the chord in direction θ is denoted by $\sigma(\theta)$ then

$$\left(\frac{1}{2\pi}\int_0^{2\pi} \sigma^2(\theta)\,d\theta\right)^{1/2} \le \frac{L(K)}{\pi}.$$

In this case, equality holds if and only if K has constant width and each chord $f(\theta)$ is a diametral chord (that is, a chord in direction θ having maximal length). Closer analysis of their argument (see their inequality (25)) shows that they proved

$$\int_0^{2\pi} \sigma^2(\theta)\,d\theta \le 2A(K+K^*), \tag{2}$$

where K^* is the reflection of K in the origin. Here equality holds if and only if each chord of the family $\mathcal{F}$ is diametral. In the same paper, Chakerian and Goodey considered the locus $M(\mathcal{F})$ of mid-points of the chords of a simple family $\mathcal{F}$. Under the assumption that this locus is piecewise smooth, they showed that

$$\int_0^{2\pi} \sigma^2(\theta)\,d\theta = 8A(K) - 8\int_{z\notin M(\mathcal{F})} w(z,M(\mathcal{F})\,dz,$$

where $w(z,M(\mathcal{F})$ is the winding number of $M(\mathcal{F})$ about z. In [8], Goodey showed that, for area and for perimeter bisectors, the midpoint locus $M(\mathcal{F})$ is traversed in the clockwise direction as the endpoints of the bisecting chords move counterclockwise. This enabled him to show that

$$\int_0^{2\pi} L^2(K,\theta)\,d\theta \ge 8A(K) \tag{3}$$

and

$$\int_0^{2\pi} A^2(K,\theta)\,d\theta \geq 8A(K). \tag{4}$$

It was also shown that equality occurs in either case only for centrally symmetric convex sets K. Returning to our example of a set described by Auerbach, we note that the mid-point locus of the area and perimeter bisectors is

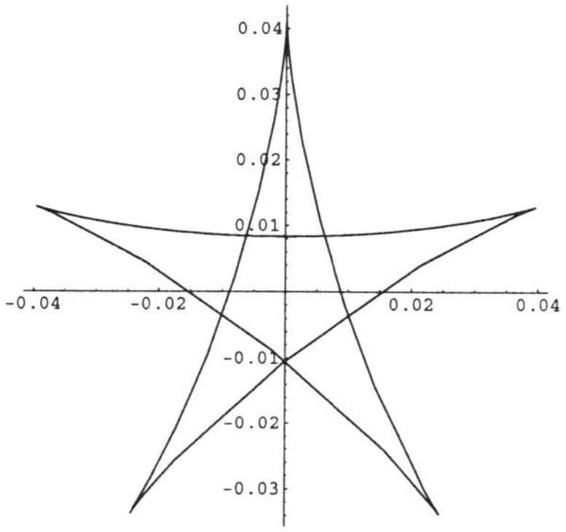

Combining the results of (2), (3) and (4), we see that

$$8A(K) \leq \int_0^{2\pi} A^2(K,\theta)\,d\theta \leq 2A(K+K^*) \tag{5}$$

and

$$8A(K) \leq \int_0^{2\pi} L^2(K,\theta)\,d\theta \leq 2A(K+K^*). \tag{6}$$

We have already noted that equality in either lower bound can only occur for centrally symmetric sets. For equality in either of the upper bounds the appropriate family of chords would have to be diametral. For equality in the right inequality of (6), the perimeter bisecting chords would have to be diametral. This implies that every shadow boundary of K divides the perimeter into parts of equal length. Schneider [15] showed that this only occurs for centrally symmetric sets. For equality in the upper bound of (5), the area bisectors would have to be diametral chords. However if the area bisecting chords are also diametral, they must then be perimeter bisecting too. Thus any equality in (5) or (6) can only occur for centrally symmetric sets. Of course, for centrally symmetric sets, all these inequalities are equalities.

References

1. H. Auerbach, Sur un probléme de M. Ulam concernant l'èquilibre des corps flottants, *Studia Math.* **7** (1938), 121–142.
2. R. Ault, Metric characterization of circles, *Amer. Math. Monthly*, **81** (1974), 149–153.
3. G. D. Chakerian, A characterization of curves of constant width, *Amer. Math. Monthly*, **81** (1974), 153–155.
4. G. D. Chakerian and P. R. Goodey, Inequalities involving convex sets and their chords, *Annals of Discrte Mathematics*, **20** (1984), 93–101.
5. Hallard T. Croft, Kenneth J. Falconer and Richard K. Guy, *Unsolved problems in geometry*, Springer-Verlag, Berlin 1991.
6. H. G. Eggleston, The maximal length of chords bisecting the area or perimeter length of plane convex sets, *Jour. London Math. Soc.* **36** (1961), 122–128.
7. P. R. Goodey, A characterization of circles, *Bull. London Math. Soc.* **4** (1972), 199–201.
8. P. R. Goodey, Mean square inequalities for chords of convex sets, *Israel Jour. Math.* **42** (1982), 132–150.
9. Hans Herda, A conjectured characterization of circles, *Amer. Math. Monthly* **78** (1971), 888–889.
10. Hans Herda, A characterization of circles and other closed curves, *Amer. Math. Monthly* **81** (1974), 146–148.
11. Erwin Lutwak, Isoperimetric inequalities involving bisectors, *Bull. London Math. Soc.,* **12** (1980), 289–295.
12. J. Pál, Ein Minimumproblem für Ovale, *Math. Ann.* **83** (1921), 311-319.
13. K. Radziszewski, Sur les cordes qui partagent l'aire d'un ovale en 2 parties égales, *Ann. Univ. Mariae Curie-Sklodowska Sect. A* **8** (1954), 89–92.
14. K. Radziszewski, Sur les cordes qui partagent le périmètre d'un ovale en 2 parties égales, *Ann. Univ. Mariae Curie-Sklodowska Sect. A* **8** (1954), 93–96.
15. R. Schneider, Über eine Integralgleichung in der Theorie der konvexen Körper, *Math. Nachr.* **44** (1970), 55–75.
16. H. S. Witsenhausen, On closed curves in Minkowski space, *Proc. Amer. Math. Soc.* **35** (1972), 240–241.
17. K. Zindler, Über konvexe Gebilde II, *Monatsh. Math. Phys.* **31** (1921), 25–27; 44–45 and 51–53.

RADON INVERSION: FROM LINES TO GRASSMANNIANS

ERIC L. GRINBERG *

Department of Mathematics
University of New Hampshire
Durham, NH 03824, USA
e-mail: grinberg@unh.edu

ABSTRACT. We begin with the earliest known inversion problems for Funk-Radon transforms and trace the development and extension of these problems as well as the ideas behind their original solutions to function spaces on Grassmann manifolds. This follows joint work with Boris Rubin. The material parallels the lecture presented at the Wuhan conference.

1. The Funk Transform on the Sphere

Some of the basic questions of integral geometry in the style of Radon and Funk are: can a function in the plane be recovered from its integrals over lines? Can a function on the sphere S^2 be recovered from its integrals over great circles?

We begin with the sphere $X = S^n \subset \mathbb{R}^n$ as our ambient space. Let $f(x)$ be a continuous function on X and let ξ be a great k-sphere in X, that is, the intersection of ξ with a k-dimensional vector subspace of $\mathbb{R}^n$. Equivalently, ξ is a *k-dimensional totally geodesic submanifold of* $\mathbb{R}^n$, or loosely, ξ is an S^k in S^n. The Funk transform is the map

$$f(x) \longrightarrow \hat{f}(\xi) \equiv \int_{x\in\xi} f(x)\, d\mu(x),$$

where $d\mu(x)$ is the rotation invariant probability measure on the k-sphere ξ. Thus we have the mapping

$$f(x) \longrightarrow \hat{f}(\xi) \quad : C(S^n) \longrightarrow C(\{S^k \subset S^n\}),$$

where the last space is the Stiefel manifold of great k-spheres in S^n. It is clear that if f is an odd function ($f(x) = -f(-x)$) then $\hat{f}$ vanishes, so it is

*Work supported in part by the National Science Foundation grant dms-9906856

natural to consider only *even* functions ($f(x) = f(-x)$), and then $\hat{f}$ can be viewed as a function of the projective k-space $[\xi] = \mathbb{R}P^k$ in the projective space $\mathbb{R}P^n$, and the Stiefel manifold of great k-spheres $\{S^k \subset S^n\}$ can now be replaced by the Grassmann manifold $G_{n,k}$ of $\mathbb{R}P^k$s in $\mathbb{R}P^n$.

2. Double Fibrations and Induced Transforms

We can summarize the above spaces and mappings in following double fibration diagram:

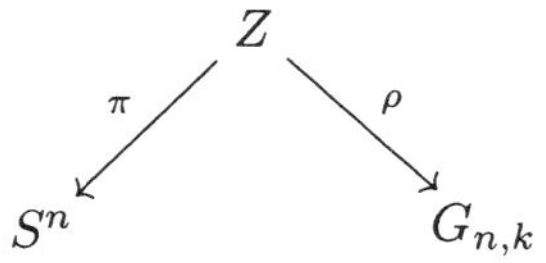

Here Z is the *incidence manifold*:

$$Z = \{(x, \xi) \,|\, x \in \xi\}.$$

The mappings ϕ and ρ are simply projections onto the first and second factors, respectively. The pull back π^* yields the following mapping of function spaces:

$$\pi^* : C(S^n) \longrightarrow C(Z),$$

while the *push forward* map, ρ_*, which can be loosely called *integration over the fiber* and which will be defined more precisely below, gives the following mapping of function spaces:

$$\rho_* : C(Z) \longrightarrow C(G_{n,k}).$$

The push forward operation acting on a k−sphere function $g(\xi)$ can be defined as follows:

$$\rho_*(g)(\xi) \equiv \int_{\{x \,|\, (x,\xi) \in Z\}} g(x, \xi) d\mu(\xi).$$

Note that the pull back ϕ^* does not require a choice of measure but ρ_* does. The Funk integral transform can be expressed now as a composition of the two mappings above:

$$\hat{f} = \rho_* \circ \pi^*(f).$$

Double fibrations were introduced in integral geometry by S.S. Chern and used in the context of integral transforms by S. Helgason, I.M. Gelfand, V. Guillemin, and many others.

In concrete examples found in the literature there is often a global group action on each ingredient of a double fibration and this makes the analysis of the induced integral transforms more tractable. Let G be a Lie group with closed subgroups K and H and consider the following *homogeneous* double fibration diagram:

$$\begin{array}{ccc} & G/K\cap H & \\ {}^{\pi}\swarrow & & \searrow^{\rho} \\ G/K & & G/H \end{array}$$

The maps π, ρ are the natural ones. For instance, the Funk transform above that integrates over S^ks in S^n can be presented by choosing:

$$G = O(n+1), \qquad K = O(n) \times \mathbb{Z}_2, \qquad H = O(k) \times O(n+1-k).$$

Here $O(\cdot)$ denotes the Orthogonal Group, the sphere S^n is presented as $O(n+1)/(O(n) \times O(1))$ (where $O(1) = \mathbb{Z}_2 = \pm 1$), and the Stiefel manifold of *great* S^ks in S^n is $O(n+1)/(O(k) \times O(n+1-k))$.

The dual transform associated with a double fibration may be obtained by composing π_* with ρ^*. In our context, if ϕ is a continuous function of S^ks in S^n, $\phi \in C(G_{n,k})$ then we define

$$\check{\phi} \equiv \pi_* \circ \rho^* \phi \equiv \int_{\{\xi \mid \xi \in x\}} \phi(\xi)\, d\mu(\xi),$$

so that

$$\phi \mapsto \check{\phi} : C(G_{n,k}) \longrightarrow C(S^n).$$

When the dimension k is even and the function f is even it is possible to invert $(\hat{f})^\vee$ by a differential operator which is a polynomial in the Laplace-Beltrami operator on S^n. See [11], [12] for real-variables derivations and [1] for a complex analytic approach to inversion on S^2.

3. Helgason's Inversion Formula

In order to obtain an inversion formula for both even and odd values of k we modify the double fibration (Helgason) as follows. Let

$$Z_p \equiv \{(x, \xi) \mid x \in S^n,\, \xi \in G_{n,k},\, \mathrm{dist}(x, \xi) = p\},$$

so that our original incidence manifold is Z_0. We now have the modified diagram

$$\begin{array}{ccc} & Z_p & \\ & {\scriptstyle\pi}\swarrow \quad \searrow{\scriptstyle\rho} & \\ S^n & & G_{n,k} \end{array}$$

We will keep the original definition of the Funk transform $f \longrightarrow \hat{f}$, but we'll modify the definition of the dual transform to read $\phi \mapsto \check{\phi}_p$, where

$$\check{\phi}_p(x) = \int_{\{\xi|\text{dist}(x,\xi)=p\}} \phi(\xi)\, d\mu(\xi).$$

We now examine the compound transform

$$f \longrightarrow \hat{f} \longrightarrow (\hat{f})_p^\vee,$$

where $(\hat{f})_p^\vee$ is defined by

$$\begin{aligned} (\hat{f})_p^\vee &\equiv \int_{\{\xi|\text{dist}(x,\xi)=p\}} \hat{f}(\xi)\, d\mu(\xi) \\ &\equiv \int_{\{\xi|\text{dist}(x,\xi)=p\}} \left(\int_{y\in\xi} f(y)\, d\mu_\xi(y) \right) d\mu(\xi). \end{aligned}$$

To transform this farther we use a bit of spherical trigonometry. If x_0 is a closest point to x in the k-sphere ξ, so that $\text{dist}(x, x_0) = p$, and y is a point in ξ, then $\text{dist}(x, y) \geq p$ and, we have the trigonometric identity

$$\cos(\text{dist}(x, y)) = \cos(\text{dist}(x, x_0)) \cdot \cos(\text{dist}(x_0, y)), \tag{1}$$

which follows because xx_0y is a right spherical triangle.

Let $r = \text{dist}(x_0, y)$ and let $q = \text{dist}(x, x_0)$. Note that, by (1), q is a function of r if p is fixed. If $f(x)$ is a function on S^n let $M^q f(x)$ be the mean value operator that averages the values of f over all points at a distance q. Then a calculation in geodesic polar coordinates yields the following alternative expression for $(\hat{f})_p^\vee$:

$$(\hat{f})_p^\vee = c_k \int_0^{\pi/2} (M^q f)\,(x)\, sin^{k-1}(r)\, dr,$$

for an appropriate constant c_k depending on the dimension k. This modified mean value operator contains a hidden form of an Abel type integral. Such integrals are inverted by a farther convolution with an appropriate kernel followed by an application of a differential operator. If $\phi(\xi)$ is a function of k-spheres ξ in S^n, let $M_v^*\phi(x)$ denote, by duality, the mean value of the function $\phi(\xi)$ over all k-spheres at a distance $\cos^{-1}(v)$ from x. Then the Helgason inversion formula for $f(x)$ can be given in the following form:

$$f(x) = c_k \Big[\Big(\frac{d}{d(u^2)}\Big)^k \int_0^u (M_v^*\varphi)(x) v^k (u^2 - v^2)^{(k-2)/2} dv\Big]_{u=1}.$$

Here $\frac{d}{d(u^2)}$ is the differential operator $\frac{1}{2u}\frac{d}{du}$ and $]_{u=1}$ denotes evaluation at $u = 1$.

4. Radon Inversion for Pairs of Grassmannians

Thus far the domain space for our integral operators has been a sphere and the range space a Grassmannian. It is natural to extend this analysis to a pair of Grassmann manifolds:

$$G_{n,k}\,,\; G_{n,k'}\;; \qquad\qquad (1 \le k < k' \le n-1).$$

The assumptions on the ranks (k, k') above follow from the usual duality and identification considerations. To extend the notion of Radon transform, we now take a function of k-planes and integrate it over all k-planes which are vector subspaces of a given k'-plane. The Radon transform of a, say, continuous function $f(\eta)$ on $G_{n,k}$ is:

$$(\mathcal{R}f)(\xi) = \int_{\eta\subset\xi} f(\eta) d_\xi \eta, \qquad \xi \in G_{n,k'}, \tag{2}$$

where $d_\xi\eta$ is the canonical normalized measure on the space of planes η in ξ. In the Funk setting it was clear that invertibility is only possible after restriction to even functions. Here we must ask for which triples (k, k', n) is it reasonable to expect inversion. General heuristics indicate that the dimension of the domain manifold should be no larger than the dimension of range manifold, so we "must" assume

$$\dim G_{n,k'} \ge \dim G_{n,k} \tag{3}$$

(otherwise, morally, the integral transform measures a smaller data set than is contained in the unknown function and hence $\mathcal{R}$ should have a non-trivial kernel). Of course, one can construct not-so-pathological injective operators from functions on a space X to functions on a space Y where $\dim X > \dim Y$, e.g., on tori by using Fourier series, or in the context of hyperbolic geometry [14].

Now $\dim G_{n,k} = k(n-k)$, so we want:

$$k + k' \leq n \qquad \text{(for } k < k'\text{)}.$$

For $k = 1$ this is a manifestation of the Funk context. In the general case there are many approaches to the inversion problem.

4.1. *Petrov's Approach*

For $k > 1$, $k' + k = n$, inversion formulae were announced by E.E Petrov in 1967 [15]. His methods involve an extension of the classical *plane waves* decomposition [13] to the domain of functions of a matrix variable. The paper [15] exhibits inversion formulae that involve divergent integrals requiring regularization. This is addressed in the sequel [16].

4.2. *The Kappa Operator Approach of Gelfand and Collaborators*

In 1967 I.M. Gelfand and collaborators introduced the Kappa Operator [3] and used it to study the Radon transform on k-dimensional planes. Later they extended this notion to the study of Radon transforms for a pair of Grassmannians [4]. The idea is to use the algebraic topology of Grassmann manifolds to exhibit inversion formulas as integrals of differential forms and to interpret the injectivity problem as a search for good homology cycles. For an introduction to this method and a variation of context see [7].

4.3. *The Lie Groups and Invariant Differential Operators Approach*

Here one uses the theory of group representations to decompose function spaces (or, more generally, spaces of sections of line bundles), so that the Radon transform becomes an intertwining operator which can be diagonalized. Inversion formulas typically can be given using invariant differential operators and range characterizations can be obtained using operators arising from the underlying Lie algebras and universal enveloping algebras. This has been studied by many authors. See, for example, [11] for inversion of the Funk transforms on spheres, [8] for projective spaces and [5], [6], [17] for Grassmannians and other symmetric spaces.

5. Inversion by Generalized Fractional Integrals

Some of the approaches above have a parity restriction: the difference of ranks $k' - k$ must be even. (The Gelfand et al approach has been extended to the odd case by means of the *Crofton symbol* [2].) Some methods require rather smooth data, e.g. C^∞ functions. The approach presented in [9] aims to take the original methods of Funk and Radon and extend them to the case involving a pair of Grassmannians. We will elaborate on this approach here.

The basic idea is to combine aspect of the underlying group of motions, fractional integral and averaging operators. Let's re-examine the Helgason inversion formula:

$$f(x) = c_k \Big[\Big(\frac{d}{d(u^2)} \Big)^k \int_0^u (M_v^* \varphi)(x) v^k (u^2 - v^2)^{(k-2)/2} dv \Big]_{u=1}.$$

Note that this formula contains a 1-dimensional Riemann-Liouville type integral. This reflects the analysis in the spherical case which reduces a general function to a zonal one, that is, a function of one variable: a function of *height*, which can be viewed as a function on the non-negative reals. In the Grassmannian case the analysis of a general function will be reduced to that of a function on the following cone of positive definite symmetric matrices:

$$\mathcal{P}_k = \text{ symm pos def } k \times k \text{ matrices.}$$

Points in $\mathcal{P}_k$ will be denoted by $r = (r_{i,j})$ or $s = (s_{i,j})$. We will also employ the generalized 'interval'

$$[0, r] \equiv \{s : \ s \in \mathcal{P}_k, \ r - s \in \mathcal{P}_k\}.$$

The preferred measure on the cone $\mathcal{P}_k$ is given by

$$ds = \prod_{i \le j} ds_{i,j}.$$

We can associate with the cone $\mathcal{P}_k$ the Siegel Gamma function

$$\Gamma_k(\alpha) = \int_{\mathcal{P}_k} e^{-\mathrm{tr}(r)} |r|^{\alpha - d} dr,$$

where $\mathrm{tr}(r)$ denotes the trace of the matrix r. One can check that integral converges absolutely for $Re\,\alpha > d-1$; it represents a product of usual Γ-functions:

$$\Gamma_k(\alpha) = \pi^{k(k-1)/4} \Gamma(\alpha) \Gamma(\alpha - \frac{1}{2}) \ldots \Gamma(\alpha - \frac{k-1}{2}).$$

There is a corresponding Beta function (C.Herz, 1955):

$$\int_0^R |r|^{\alpha-d}|R-r|^{\beta-d}dr = B_k(\alpha,\beta)|R|^{\alpha+\beta-d},$$

where

$$B_k(\alpha,\beta) = \frac{\Gamma_k(\alpha)\,\Gamma_k(\beta)}{\Gamma_k(\alpha+\beta)};$$

here the integration is over the interval $[0,r]$ defined above and

$$Re\,\alpha, Re\,\beta > d-1; \qquad R \in \mathcal{P}_k.$$

The Riemann-Liouville integral in the Helgason inversion formula will be replaced by the following higher rank counterpart:

$$(I_+^\alpha w)(r) = \frac{1}{\Gamma_k(\alpha)} \int_0^r w(s)\,(\det(r-s))^{\alpha-(k+1)/2}ds,$$

where $\det(r-s)$ is the determinant of the matrix $r-s$ and we assume $(Re\,\alpha > (k-1)/2)$. Integrals of this type occur in the works of L. Gårding, M. Riesz, C.L. Siegel, S. Bochner, S. Gindikin and others.

We will also need a differential operator on our function spaces which will play a role in inversion formulas. Let

$$\eta_{i,j} = \begin{cases} 1 & \text{if } i=j \\ 1/2 & \text{if } i \neq j. \end{cases}$$

Given a function $f(r)$, $r=(r_{i,j}) \in \mathcal{P}_k$, we define

$$(D_+ f)(r) = \det\left(\eta_{i,j}\,\frac{\partial}{\partial r_{i,j}}\right) f(r).$$

It follows that $D_+ I_+^\alpha = I_+^{\alpha-1}$. This is an extension of a variant of a familiar identity for Riesz potentials and is due to Gårding.

Another important ingredient in spherical inversion is the mean value $(M^*_{\cos^{-1}(v)}\varphi)(x)$. This is the average of the k'-plane function $\varphi(\xi)$ over the set of all $\xi \in G_{n,k'}$ satisfying $\cos\theta = v$, where θ is the angle between the unit vector x and the *orthogonal projection* $\mathrm{Pr}_\xi x$ of x onto ξ. This leads to a generalization.

Given a k-plane $\eta \in G_{n,k}$, let $\eta^\perp$ be the $(n-k)$-subspace orthogonal to η. Let $\tau \in G_{n,\ell}$ be an ℓ-plane, with $\ell \leq k$. Suppose that τ is spanned by the orthogonal $n \times l$ matrix y, with columns $y_1, \ldots, y_\ell$. Define

$$\mathrm{Cos}^2(\eta,\tau) = y'\mathrm{Pr}_\eta y, \qquad \mathrm{Sin}^2(\eta,\tau) = y'\mathrm{Pr}_{\eta^\perp} y,$$

where y' is transpose of y.

Both quantities are independent of the choice of τ-basis $y = [y_1, \ldots, y_\ell]$ in τ, and both represent positive semidefinite $\ell \times \ell$ matrices. (Replace the linear operator Pr_η by the matrix xx' where $x = [x_1, \ldots, x_k]$ is an orthonormal η-basis.) Clearly,

$$Cos^2(\eta, \tau) + \mathrm{Sin}^2(\eta, \tau) = I_\ell \quad \text{(the identity matrix).}$$

We now introduce the matrix analogue of $(M^*_{\cos^{-1}(v)}\varphi)(x)$:

$$(M^*_r\varphi)(\eta) = \int\limits_{\{\xi:\, \mathrm{Cos}^2(\xi,\eta)=r\}} \varphi(\xi) dm_\eta(\xi).$$

Here $\eta \in G_{n,k}, \xi \in G_{n,k'}, r \in \mathcal{P}_k$, and $dm_\eta(\xi)$ is the relevant normalized measure. The average is over positive definite matrices $r \in \mathcal{P}_k$. With these ingredients in hand we can present an inversion formula involving a pair of Grassmannians. In reading the theorem below one can safely assume that the function f (the "data") is continuous.

Theorem ([9]). *Let f be an L^p function on the Grassmannian $G_{n,k}$ with $1 \le p \le \infty$. Let $G_{n,k'}$ be a companion Grassmannian, where k' satisfies*

$$1 \le k < k' \le n - 1, \quad k + k' \le n.$$

Let $\varphi(\xi)$ be the function on $G_{n,k'}$ which is the Radon transform of f.

$$1 \le k < k' \le n - 1, \quad k + k' \le n.$$

Denote the semi rank difference $(k' - k)/2$ by α and the normalization constant $\frac{\Gamma_\ell(k'/2)}{\Gamma_\ell(k/2)}$ by c. Let $\hat{\varphi}_\eta(r)$ be the rescaled mean value of the function ϕ:

$$\hat{\varphi}_\eta(r) = (\det(r))^{\alpha - 1/2}(M^*_r\varphi)(\eta).$$

Then for any integer $m > (k' - 1)/2$ we have the inversion formula

$$f(\eta) = c^{-1} \lim_{r \to I_k} (D^m_+ I^{m-\alpha}_+ \hat{\varphi}_\eta)(r).$$

Here differentiation is understood in the sense of distributions. In case the where the semi rank difference $(k' - k)/2$ is a positive integer ℓ, we have the simplified formula

$$f = c^{-1} \lim_{r \to I_k} (D^\ell_+ \hat{\varphi}_\eta)(r).$$

In the inversion formulae limits the symmetric matrix r approaches the $k \times k$ identity matrix and the limit is in the L^p sense.

It is natural to consider the Radon inversion problem for Grassmannians over other division rings, e.g., complex, quaternionic and octonionic, as well as transforms attached to specialized Grassmannians such as isotropic spaces for nondegenerate quadratic forms. For the first two cases an extension was recently obtained by Genkai Zhang [18].

References

1. Bailey, T. N.; Eastwood, M. G.; Gover, A. R.; Mason, L. J *Complex analysis and the Funk transform* J. Korean Math. Soc., no. 40, (2003), no. 4, 577–593.
2. Gel'fand, I.M., Graev, M.I., and Rosu, R., *The problem of integral geometry and intertwining operators for a pair of real Grassmannian manifolds*, J. Operator Theory, no. 12, (1984), 339–383.
3. Gel'fand, I.M., Graev, M.I., and Šapiro, Z.Ja., *Integral geometry on k-dimensional planes*, Funct. Anal. Appl., no. 1, (1967), 14-27.
4. Gel'fand, I.M., Graev, M.I., and Šapiro, Z.Ja. , *A problem of integral geometry connected with a pair of Grassmann manifolds*, Dokl. Akad. Nauk SSSR, **193**, no. 2, (1970), 892-896.
5. Gonzalez, F.B. and Kakehi, T., *Pfaffian systems and Radon transforms on affine Grassmann manifolds*, Math. Ann. 326 (2003), no. 2, 237–273.
6. Grinberg, E.L. *Radon transforms on higher rank Grassmannians*, J. Differential Geometry **24** (1986), 53-68.
7. ——— *That Kappa operator: Gel'fand-Graev-Shapiro inversion and Radon transforms on isotropic planes*, Tomography, impedance imaging, and integral geometry (South Hadley, MA, 1993), 93–104.
8. ——— *Spherical harmonics and integral geometry on projective spaces* Trans. Amer. Math. Soc. 279 (1983), no. 1, 187–203.
9. Grinberg, E.L. and Rubin, B., *Radon inversion on Grassmannians via Gårding-Gindikin fractional integrals*, Ann. of Math.(2) 159 (2004), no. 2, 783–817
10. Guillemin, V. and Sternberg, S. *Geometric Asymptotics*, rev. Ed. Amer. Math. Soc. (1990)
11. Guillemin, V. *The Radon transform on Zoll sufaces*, Adv. Math. 22 (1976), 85-119.
12. Helgason, S. *The Radon Transform* Birkhäuser, Boston, Second edition, (1999).
13. John, F. *Plane Waves and Spherical Means Applied to Partial Differenial Equations*, Dover Publications, New York, (2004).
14. Lax, P., private communication.
15. Petrov, E.E., *The Radon transform in spaces of matrices and in Grassmann manifolds*, Dokl. Akad. Nauk SSSR, **177**, No. 4 (1967), 1504-1507.
16. ——— *The Radon transform in spaces of matrices*, Trudy seminara po vek-

tornomu i tenzornomu analizu, M.G.U., Moscow, **15** (1970), 279-315 (Russian).
17. Rouvire, F. *Inverting Radon transforms: the group-theoretic approach*, Enseign. Math. (2) 47 (2001), no. 3-4, 205–252.
18. Zhang, Genkai, *Radon Transform on Real, Complex and Quaternionic Grassmannians*, Preprint (2005).

VALUATIONS IN THE AFFINE GEOMETRY OF CONVEX BODIES

MONIKA LUDWIG
Institut für Diskrete Mathematik und Geometrie
Technische Universität Wien
Wiedner Hauptstraße 8-10/104
1040 Wien, Austria
E-mail: monika.ludwig@tuwien.ac.at

A survey on SL(n) invariant and SL(n) covariant valuations is given.

1. Introduction

A function Φ defined on convex bodies in $\mathbb{R}^n$ and taking values in an Abelian semigroup is called a *valuation* if

$$\Phi(K) + \Phi(L) = \Phi(K \cup L) + \Phi(K \cap L) \quad \text{for } K, L, K \cup L \in \mathcal{K}^n,$$

where $\mathcal{K}^n$ is the set of convex bodies (convex, compact sets) in $\mathbb{R}^n$. Thus the notion of valuation is a generalization of the notion of measure. In the 1930s, Blaschke obtained the first classification of real valued valuations that are SL(n) invariant. This was greatly extended by Hadwiger in his famous classification of continuous and rigid motion invariant valuations.

Theorem 1 (Hadwiger [26]). *A functional $\Phi : \mathcal{K}^n \to \mathbb{R}$ is a continuous and rigid motion invariant valuation if and only if there are constants c_0, $c_1, \ldots, c_n \in \mathbb{R}$ such that*

$$\Phi(K) = c_0\, V_0(K) + \ldots + c_n\, V_n(K)$$

for every $K \in \mathcal{K}^n$.

Here $V_0(K), \ldots, V_n(K)$ are the intrinsic volumes of K; $V_n = V$ is the ordinary volume, V_{n-1} is proportional to the surface area and V_0 is the Euler Characteristic. The classical theory of valuations and their applications in integral geometry and geometric probability are described in the books and surveys [26,35,67,70].

In recent years there have been many new developments in the theory of valuations (see [1,7,12,13,19,32,34,68,69,82,84]). Here we confine our attention to valuations in the affine geometry of convex bodies. Let SL(n) denote the special linear group, that is, the group of $n \times n$ matrices of determinant 1. We say that a functional Φ is SL(n) *invariant* if

$$\Phi(\alpha K) = \Phi(K) \qquad \forall K \in \mathcal{K}^n, \forall \alpha \in \mathrm{SL}(n).$$

We say that Φ is *equi-affine invariant* if it is SL(n) invariant and translation invariant. These notions are important for real valued valuations. We say that a functional Φ is SL(n) *covariant* if

$$\Phi(\alpha K) = \alpha\, \Phi(K) \qquad \forall K \in \mathcal{K}^n, \forall \alpha \in \mathrm{SL}(n).$$

This notion is important for vector and tensor valued valuations as well as for convex body and star body valued valuations. In the following, we describe classification theorems for SL(n) invariant and SL(n) covariant valuations and make some remarks on related results for rotation invariant and rotation covariant valuations.

2. Real valued valuations on polytopes

Clearly, the intrinsic volumes $V_1, \dots, V_{n-1}$ are not SL(n) invariant and Hadwiger's characterization theorem implies that every continuous, equi-affine invariant valuation on $\mathcal{K}^n$ is a linear combination of V_0 and V_n. It is very easy to see this directly. We sketch this well-known proof here. The result is far from best possible (there are characterizations of volume using only translation invariance, see [67] and [32]) but the arguments are used in one form or another in many of the proofs for characterization theorems.

Let $\mathcal{P}^n$ denote the set of convex polytopes in $\mathbb{R}^n$. Let $\Phi : \mathcal{P}^n \to \mathbb{R}$ be an equi-affine invariant valuation. Since Φ is translation invariant, we have $\Phi(\{x\}) = c_0$ for every singleton $\{x\}$, $x \in \mathbb{R}^n$. Therefore the functional $\Phi_0(P) = \Phi(P) - c_0$ is *simple*, that is, it vanishes on lower dimensional sets. Now let S be an n-dimensional simplex of volume s. Since Φ_0 is SL(n) invariant, $\Phi_0(S)$ depends only on s, that is, there is a function $f : [0, \infty] \to \mathbb{R}$ such that $\Phi_0(S) = f(s)$. We can subdivide S by cutting with a hyperplane containing an $(n-2)$-dimensional face of S into two simplices S_1, S_2 of volume s_1 and s_2, respectively. Since Φ_0 is a simple valuation, we have $\Phi_0(S) = \Phi_0(S_1) + \Phi_0(S_2)$ and therefore

$$f(s) = f(s_1 + s_2) = f(s_1) + f(s_2).$$

This holds for every $s_1, s_2 \geq 0$. Thus f is a solution of Cauchy's functional equation. If we assume that Φ is (Borel) measurable, we can conclude that $f(s) = c_1\, s$. Thus

$$\Phi(P) = c_0 + c_1\, V(P)$$

for every $P \in \mathcal{P}^n$.

Next, we consider the corresponding problem on $\mathcal{P}_0^n$, the set of convex polytopes that contain the origin in their interiors. Here the situation is not yet well understood. It is easy to see that on $\mathcal{P}_0^n$ there are additional examples of $\mathrm{SL}(n)$ invariant valuations. We describe the construction since it will also be used for tensor, convex body and star body valued valuations. For $\Phi : \mathcal{P}_0^n \to \mathbb{R}$ an $\mathrm{SL}(n)$ invariant valuation, set $\Psi(P) = \Phi(P^*)$. Here P^* is the polar body of $P \in \mathcal{P}_0^n$, that is,

$$P^* = \{y \in \mathbb{R}^n \,|\, x \cdot y \leq 1 \text{ for all } x \in P\}$$

and $x \cdot y$ denotes the inner product x and y in $\mathbb{R}^n$. The functional $\Psi : \mathcal{P}_0^n \to \mathbb{R}$ has the following properties. For $P, Q, P \cup Q \in \mathcal{P}_0^n$, we have

$$(P \cup Q)^* = P^* \cap Q^* \text{ and } (P \cap Q)^* = P^* \cup Q^*.$$

Since Φ is a valuation,

$$\begin{aligned} \Psi(P) + \Psi(Q) = \quad & \Phi(P^*) + \Phi(Q^*) = \\ & \Phi(P^* \cup Q^*) + \Phi(P^* \cap Q^*) = \\ & \Phi((P \cap Q)^*) + \Phi((P \cup Q)^*) = \Psi(P \cap Q) + \Psi(P \cup Q), \end{aligned}$$

that is, Ψ is also a valuation. For $\alpha \in \mathrm{SL}(n)$ and $P \in \mathcal{P}_0^n$, we have

$$(\alpha\, P)^* = \alpha^{-t}\, P^*,$$

where α^{-t} is the inverse of the transpose of α. Since Φ is $\mathrm{SL}(n)$ invariant,

$$\Psi(\alpha P) = \Phi((\alpha P)^*) = \Phi(\alpha^{-t} P^*) = \Phi(P^*) = \Psi(P),$$

that is, Ψ is also $\mathrm{SL}(n)$ invariant. We say that a functional Φ is *homogeneous of degree* q if

$$\Phi(t\, K) = t^q \Phi(K) \qquad \forall K \in \mathcal{K}^n, \forall t \geq 0.$$

If Φ is homogeneous of degree q, then

$$\Psi(t\, P) = \Phi((t\, P)^*) = \Phi(t^{-1} P^*) = t^{-q}\, \Psi(P),$$

that is, Ψ is homogeneous of degree $-q$. In particular, this shows that $K \mapsto V(K^*)$ is an $\mathrm{SL}(n)$ invariant and homogeneous valuation. The next result shows that there are no further examples.

Theorem 2 ([46]). *A functional $\Phi : \mathcal{P}_0^n \to \mathbb{R}$ is a measurable,* SL(n) *invariant valuation which is homogeneous of degree q if and only if there is a constant $c \in \mathbb{R}$ such that*

$$\Phi(P) = \begin{cases} c & \text{for } q = 0 \\ c\,V(P) & \text{for } q = n \\ c\,V(P^*) & \text{for } q = -n \\ 0 & \text{otherwise} \end{cases}$$

for every $P \in \mathcal{P}_0^n$.

It is not known if there are additional examples if Φ is not homogeneous. We conjecture that every SL(n) invariant and continuous valuation is a linear combination of a constant, the volume of the body and the volume of the polar body. Also the problem to classify rotation invariant valuations on $\mathcal{P}_0^n$ is open. Alesker [1] has obtained a classification of continuous, rotation invariant, polynomial valuations on $\mathcal{K}^n$.

3. Real valued valuations on convex bodies

There are SL(n) invariant valuations on convex bodies that vanish on polytopes. The *affine surface area* $\Omega : \mathcal{K}^n \to \mathbb{R}$, is such a functional. It is defined by

$$\Omega(K) = \int_{\partial K} \kappa(K,x)^{1/(n+1)}\,dx,$$

where $\kappa(K,x)$ is the generalized Gaussian curvature of K at x and ∂K is the boundary of K. Affine surface area was introduced by Pick and Blaschke at the beginning of the twentieth century in the context of Affine Differential Geometry. In the 1990s, it has been extended to a functional for general (not necessarily smooth) convex bodies by Leichtweiß, Lutwak, Schütt and Werner (see [39]). Affine surface area is an equi-affine invariant valuation. Lutwak [56] proved that Ω is upper semicontinuous.

Affine surface area has found a wide field of applications. In particular, affine surface area describes the quality of polytopal volume approximation (see [24,41,80,86]). See also [89,90,91]. In the planar case, there is a nice geometric interpretation for affine surface area, see [10]. In general dimensions, floating bodies (see [39]) and related constructions (see [71,92,93]) are used to obtain geometric interpretations. There is the following characterization of Ω.

Theorem 3 ([42,51]). *A functional $\Phi : \mathcal{K}^n \to \mathbb{R}$ is an upper semicontinuous and equi-affine invariant valuation if and only if there are constants c_0, c_1,* *and* $c_2 \geq 0$ *such that*

$$\Phi(K) = c_0 + c_1 V(K) + c_2 \Omega(K)$$

for every $K \in \mathcal{K}^n$.

This is an equi-affine analogue of Hadwiger's Characterization Theorem. The problem to determine all upper semicontinuous and rigid motion invariant valuations on $\mathcal{K}^n$ is open. Only in the planar case, there is a complete classification (see [43]).

Let $\mathcal{K}_0^n$ denote the space of convex bodies that contain the origin in their interiors. A classical notion is the *centro-affine surface area* $\Omega_c(K)$ that can be defined in the following way. If $K \in \mathcal{K}_0^n$ and the Gaussian curvature $\kappa(K, x)$ exists for $x \in \partial K$, set

$$\kappa_0(K, x) = \frac{\kappa(K, x)}{(x \cdot u(K, x))^{n+1}},$$

where $u(K, x)$ is the exterior normal unit vector to K at $x \in \partial K$. Note that $\kappa_0(K, x)^{-1/2}$ is (up to a constant) the volume of the centered ellipsoid that osculates K at x. Let

$$d\sigma_K(x) = (x \cdot u(K, x))\, dx$$

denote the cone measure of K. Then the centro-affine surface area is defined by

$$\Omega_c(K) = \int_{\partial K} \kappa_0(K, x)^{\frac{1}{2}}\, d\sigma_K(x)$$

and it is $\mathrm{GL}(n)$ invariant. Here $\mathrm{GL}(n)$ denotes the general linear group, that is, the group of invertible $n \times n$ matrices. Similar to affine surface area, centro-affine surface area has applications in polytopal approximation. In particular, centro-affine surface area describes the quality of polytopal approximation with respect to the Banach-Mazur distance (see [23]). There is the following characterization of Ω_c.

Theorem 4 ([52]). *A functional* $\Phi : \mathcal{K}_0^n \to \mathbb{R}$ *is an upper semicontinuous,* $\mathrm{GL}(n)$ *invariant valuation if and only if there are constants* $c_0 \in \mathbb{R}$ *and* $c_1 \geq 0$ *such that*

$$\Phi(K) = c_0 + c_1 \Omega_c(K)$$

for every $K \in \mathcal{K}_0^n$.

More generally, the following classification of $\mathrm{SL}(n)$ invariant and upper semicontinuous valuations holds.

Theorem 5 ([52]). *A functional $\Phi : \mathcal{K}_0^n \to \mathbb{R}$ is an upper semicontinuous and* SL(n) *invariant valuation that vanishes on $\mathcal{P}_0^n$ if and only if there is a concave function $\phi : [0,\infty) \to [0,\infty)$ with $\lim_{t\to 0} \phi(t) = 0$ and $\lim_{t\to\infty} \phi(t)/t = 0$ such that*

$$\Phi(K) = \int_{\partial K} \phi(\kappa_0(K,x))\, d\sigma_K(x)$$

for every $K \in \mathcal{K}_0^n$.

Combined with Theorem 2 this gives a classification of upper semicontinuous, SL(n) invariant, homogeneous valuations on $\mathcal{K}_0^n$.

Theorem 6 ([52]). *A functional $\Phi : \mathcal{K}_0^n \to \mathbb{R}$ is an upper semicontinuous,* SL(n) *invariant valuation that is homogeneous of degree q if and only if there are constants $c_0 \in \mathbb{R}$ and $c_1 \geq 0$ such that*

$$\Phi(K) = \begin{cases} c_0 + c_1\, \Omega_n(K) & \text{for } q = 0 \\ c_1\, \Omega_p(K) & \text{for } -n < q < n,\ q \neq 0, \\ c_0\, V(K) & \text{for } q = n \\ c_0\, V(K^*) & \text{for } q = -n \\ 0 & \text{otherwise} \end{cases}$$

for every $K \in \mathcal{K}_0^n$ where $p = n(n-q)/(n+q)$.

Here $\Omega_p(K)$ is the *L_p-affine surface area* of K. This notion was introduced by Lutwak [60] within the setting of the L_p-Brunn-Minkowski theory. For $K \in \mathcal{K}_0^n$ and $p \geq 1$, he defined

$$\Omega_p(K) = \int_{S^{n-1}} f_p(K,u)^{\frac{n}{n+p}}\, du$$

where S^{n-1} is the unit sphere and $f_p(K,\cdot)$ is the L_p-curvature function of K. Lutwak [60] showed that Ω_p is SL(n) invariant and upper semicontinuous on $\mathcal{K}_0^n$. Hug [28] gave an equivalent definition of L_p-affine surface areas and extended Lutwak's definition from $p \geq 1$ to $p > 0$. Hug's definition can be written in the following way. For $p > 0$ and $K \in \mathcal{K}_0^n$,

$$\Omega_p(K) = \int_{\partial K} \kappa_0(K,x)^{\frac{p}{n+p}}\, d\sigma_K(x).$$

For $p = 1$ we obtain the affine surface area $\Omega(K)$ and for $p = n$ the centro-affine surface area $\Omega_c(K)$. Geometric interpretations of L_p-affine surface areas can be found in [72,87] and applications to partial differential equations in [61].

There is the following application of Theorem 6. Set $\Psi(K) = \Omega_p(K^*)$. Then Ψ is also an $\mathrm{SL}(n)$ invariant valuation and it is homogeneous of degree $-q = -n\,(n-p)/(n+p)$. Therefore by Theorem 6, there is a constant $c \geq 0$ such that $\Psi(K) = c\,\Omega_r(K)$ where $r = n^2/p$. Since for the ball B of radius 1 all L_p-affine surface areas coincide, we have $\Omega_p(B) = \Omega_p(B^*) = c\,\Omega_r(B)$ and $c = 1$. Thus

$$\Omega_p(K^*) = \Omega_{n^2/p}(K).$$

This result was obtained by Hug [29] using a different approach.

4. Tensor valued valuations

For vector valued valuations, Schneider [78] proved the following analogue of Hadwiger's characterization theorem: Every continuous, rotation covariant, vector valued valuation z on $\mathcal{K}^n$ with the property that $z(K+x) - z(K)$ is parallel to x for every $x \in \mathbb{R}^n$ is a linear combination of quermassvectors (see also [27] and [81], Chapter 5.4). Here we are interested in $\mathrm{SL}(n)$ covariant vector valued valuations on $\mathcal{P}_0^n$. For this question, the fundamental notion is the *moment vector*

$$m(P) = \int_P x\,dx$$

of $P \in \mathcal{P}_0^n$, that is, $m(P)$ is the centroid of P multiplied by the volume of P. The moment vector is an $\mathrm{SL}(n)$ covariant valuation on $\mathcal{P}_0^n$. The problem to classify $\mathrm{SL}(n)$ covariant valuations from $\mathcal{P}_0^n$ to $\mathbb{R}^n$ is not completely solved but there is a classification of $\mathrm{GL}(n)$ covariant valuations. Here a function $z : \mathcal{P}_0^n \to \mathbb{R}^n$ is called $\mathrm{GL}(n)$ *covariant*, if there is a real number q such that

$$z(\alpha P) = |\det \alpha|^q\, \alpha z(P) \qquad \forall P \in \mathcal{P}_0^n, \forall \alpha \in \mathrm{GL}(n).$$

A function is (Borel) measurable if the pre-image of every open set is a Borel set.

Theorem 7 ([44]). *A function $z : \mathcal{P}_0^n \to \mathbb{R}^n$, $n \geq 3$, is a $\mathrm{GL}(n)$ covariant, measurable valuation if and only if there is a constant $c \in \mathbb{R}$ such that*

$$z(P) = c\, m(P)$$

for every $P \in \mathcal{P}_0^n$.

More general tensor valued valuations on $\mathcal{K}^n$ were studied and classified in [2,68,69,83,84]. Here we consider only symmetric tensors of rank 2, that is, functions $Z : \mathcal{P}_0^n \to \mathcal{M}^n$, where $\mathcal{M}^n$ is the set of real symmetric $n \times n$ matrices. Note that to every positive definite matrix $A \in \mathcal{M}^n$ corresponds an ellipsoid E_A defined by

$$E_A = \{x \in \mathbb{R}^n : x \cdot Ax \leq 1\}. \tag{1}$$

A classical concept from mechanics is the *Legendre ellipsoid* or *ellipsoid of inertia* $\Gamma_2 K$ associated with a convex body $K \subset \mathbb{R}^n$ (see [39,40,73]). It can be defined as the unique ellipsoid centered at the center of mass of K such that the ellipsoid's moment of inertia about any axis passing through the center of mass is the same as that of K. The Legendre ellipsoid can also be defined by the *moment matrix* $M_2(K)$ of K. This is the $n \times n$ matrix with coefficients

$$\int_K x_i \, x_j \, dx,$$

where we use coordinates $x = (x_1, \ldots, x_n)$ for $x \in \mathbb{R}^n$. For a convex body K with non-empty interior, $M_2(K)$ is a positive definite symmetric $n \times n$ matrix and using (1) we have

$$\Gamma_2 K = \sqrt{\frac{n+2}{V(K)}}\, E_{M_2(K)^{-1}}.$$

Note that $M_2 : \mathcal{K}^n \to \mathcal{M}^n$ is $\mathrm{GL}(n)$ covariant of weight $q = 1$, where a function $Z : \mathcal{P}_0^n \to \mathcal{M}^n$ is $\mathrm{GL}(n)$ covariant if there is a real number q such that

$$Z(\alpha P) = |\det \alpha|^q \, \alpha Z(P) \alpha^t \qquad \forall K \in \mathcal{K}^n, \forall \alpha \in \mathrm{GL}(n).$$

Here α^t denotes the transpose of α. There is the following classification of $\mathrm{GL}(n)$ covariant matrix valued valuations.

Theorem 8 ([47]). *A function $Z : \mathcal{P}_0^n \to \mathcal{M}^n$, $n \geq 3$, is a measurable, $\mathrm{GL}(n)$ covariant valuation if and only if there is a constant $c \in \mathbb{R}$ such that*

$$Z(P) = c\, M_2(P) \quad \text{or} \quad Z(P) = c\, M_{-2}(P^*)$$

for every $P \in \mathcal{P}_0^n$.

Here $M_{-2}(P^*)$ is the matrix with coefficients

$$\sum_u \frac{a(P^*, u)}{h(P^*, u)}\, u_i \, u_j$$

where the sum is taken over all unit normals u of facets of P^* and where $a(P^*, u)$ is the $(n-1)$-dimensional volume of the facet with normal u and

$h(P^*, u)$ is the distance from the origin of the hyperplane containing this facet. This matrix corresponds to the ellipsoid $\Gamma_{-2}P^*$ recently introduced Lutwak, Yang, and Zhang [63]. Using (1), this *LYZ ellipsoid* is given by

$$\Gamma_{-2}P^* = \sqrt{V(K)}\, E_{M_{-2}(P^*)}.$$

More information on this ellipsoid, its applications, and its connection to the Fisher information from information theory can be found in [25,63,65].

5. Convex body valued valuations

The basic notion of addition for convex bodies is Minkowski addition. For $K_1, K_2 \in \mathcal{K}^n$, the Minkowski sum is

$$K_1 + K_2 = \{x_1 + x_2 : x_1 \in K_1, x_2 \in K_2\}$$

and $K_1 + K_2 \in \mathcal{K}^n$. Minkowski addition can also be described by using the *support function* $h(K, \cdot)$, which is defined for $u \in S^{n-1}$ by

$$h(K, u) = \max\{x \cdot u : x \in K\}.$$

Note that $h(K, \cdot)$ on S^{n-1} determines K and that the support function of the Minkowski sum is given by

$$h(K_1 + K_2, \cdot) = h(K_1, \cdot) + h(K_2, \cdot).$$

Minkowski addition and volume are the fundamental notions in the Brunn-Minkowski theory (see [81]). We remark that there are important extensions of the concepts of the Brunn-Minkowski theory in the L_p-Brunn-Minkowski theory (see [57,60]).

Here we consider convex body valued functions on $\mathcal{K}^n$ and $\mathcal{K}_0^n$ that are valuations with respect to Minkowski addition. Since we are interested in the affine geometry of convex bodies, we confine our attention to operators $\mathrm{Z} : \mathcal{K}^n \to \mathcal{K}^n$ that are $\mathrm{SL}(n)$ covariant or $\mathrm{SL}(n)$ contravariant. Here an operator is called $\mathrm{SL}(n)$ *contravariant* if

$$\mathrm{Z}(\alpha K) = \alpha^{-t}\, \mathrm{Z}\, K \qquad \forall K \in \mathcal{K}_0^n, \forall \alpha \in \mathrm{SL}(n),$$

where α^{-t} is the transpose of the inverse of α.

The classical example of an $\mathrm{SL}(n)$ contravariant operator is the *projection operator* $\Pi : \mathcal{K}^n \to \mathcal{K}^n$. It is defined in the following way. The *projection body*, ΠK, of K is the convex body whose support function is given by

$$h(\Pi K, u) = \mathrm{vol}(K|u^\perp) \quad \text{for } u \in S^{n-1},$$

where vol denotes $(n-1)$-dimensional volume and $K|u^\perp$ denotes the image of the orthogonal projection of K onto the subspace orthogonal to u. Projection bodies were introduced by Minkowski at the turn of the last century. They are an important tool for studying projections. Petty [75] showed that

$$\Pi(\alpha K) = |\det \alpha|\, \alpha^{-t}\Pi K \quad \text{and} \quad \Pi(K+x) = \Pi K \tag{2}$$

for every $K \in \mathcal{K}^n$, $\alpha \in \mathrm{GL}(n)$, and $x \in \mathbb{R}^n$. It follows from (2) that the volume of ΠK and of the polar of ΠK are affine invariants, and there are important affine isoperimetric inequalities for these quantities (see [76,94,58,20,64,96]). There is the following characterization of Π.

Theorem 9 ([45,48]). *An operator $Z : \mathcal{P}^n \to \mathcal{K}^n$ is an* SL(n) *contravariant and translation invariant valuation if and only if there is a constant $c \geq 0$ such that*

$$ZP = c\,\Pi P$$

for every $P \in \mathcal{P}^n$.

A simple consequence of this characterization is that every continuous, SL(n) contravariant, translation invariant valuation on $\mathcal{K}^n$ is a multiple of the projection operator.

The corresponding result for SL(n) covariant operators is the following.

Theorem 10 ([48]). *An operator $Z : \mathcal{P}^n \to \mathcal{K}^n$ is an* SL(n) *covariant and translation invariant valuation if and only if there is a constant $c \geq 0$ such that*

$$ZP = c\,\mathrm{D}\,P$$

for every $P \in \mathcal{P}^n$.

Here $\mathrm{D}\,P = P + (-P)$ is the *difference body* of P, which is an important concept in the affine geometry of convex bodies. The fundamental affine isoperimetric inequality for difference bodies is the Rogers-Shephard inequality [77].

It is an open problem to establish a classification of rigid motion covariant convex body valued valuations. But there are some important results. An operator $\mathrm{Z} : \mathcal{K}^n \to \mathcal{K}^n$ is *Minkowski additive* if $\mathrm{Z}(K_1+K_2) = \mathrm{Z}\,K_1+\mathrm{Z}\,K_2$ for $K_1, K_2 \in \mathcal{K}^n$. Note that every Minkowski additive operator is a valuation with respect to Minkowski addition but not vice versa. Continuous Minkowski additive operators that commute with rigid motions are called endomorphisms. Schneider [79] (see also [81]) showed that there is a great variety of these operators. He obtained a complete classification of endomorphisms in $\mathcal{K}^2$ and characterizations of special endomorphisms in $\mathcal{K}^n$. These results were further extended by Kiderlen [31]. Also operators that map Blaschke sums of convex bodies to Minkowski sums are examples of valuations with respect to Minkowski addition. For these operators, classification results were obtained by Schuster [85].

Next, we consider operators on $\mathrm{Z} : \mathcal{K}_0^n \to \mathcal{K}^n$. Such an operator is called $\mathrm{GL}(n)$ *covariant*, if there is a real number q such that

$$\mathrm{Z}(\alpha K) = |\det \alpha|^q \, \alpha \, \mathrm{Z}\, K \qquad \forall K \in \mathcal{K}_0^n, \forall \alpha \in \mathrm{GL}(n).$$

It is called $\mathrm{GL}(n)$ *contravariant*, if there is a real number q such that

$$\mathrm{Z}(\alpha K) = |\det \alpha|^q \, \alpha^{-t} \, \mathrm{Z}\, K \qquad \forall K \in \mathcal{K}_0^n, \forall \alpha \in \mathrm{GL}(n).$$

Note that the projection operator is $\mathrm{GL}(n)$ contravariant of weight $q = 1$ and that the operator $K \mapsto \Pi K^*$ is $\mathrm{GL}(n)$ covariant of weight $q = -1$. Further examples of $\mathrm{GL}(n)$ covariant operators are the *trivial* operators $K \mapsto c_0 K + c_1(-K)$, $c_0, c_1 \geq 0$.

Theorem 11 ([48,50]). *An operator* $\mathrm{Z} : \mathcal{P}_0^n \to \mathcal{K}^n$ *is a non-trivial* $\mathrm{GL}(n)$ *covariant valuation if and only if there are constants* $c_0 \geq 0$ *and* $c_1 \in \mathbb{R}$ *such that*

$$\mathrm{Z}\, P = c_0 \, \mathrm{M}\, P + c_1 \, m(P) \quad \text{or} \quad \mathrm{Z}\, P = c_0 \, \Pi \, P^*$$

for every $P \in \mathcal{P}_0^n$.

Here $\mathrm{M}\, P$ is the *moment body* of $P \in \mathcal{P}_0^n$, that is, the convex body whose support function is given by

$$h(\mathrm{M}\, K, u) = \int_K |u \cdot x| \, dx \quad \text{for } u \in S^{n-1}.$$

If the n-dimensional volume $V(K)$ of K is positive, then the *centroid body* ΓK of K is defined by

$$\Gamma K = \frac{1}{V(K)} \mathrm{M}\, K.$$

Centroid bodies are a classical notion from geometry (see [16,39,81]). If K is centrally symmetric, then ΓK is the body whose boundary consists of the locus of the centroids of the halves of K formed when K is cut by hyperplanes through the origin. The fundamental affine isoperimetric inequality for centroid bodies is the Busemann-Petty centroid inequality [74]. Recent results on centroid bodies can be found in [11,17,22,53,55,62,66,73].

6. Star body valued valuations

The basic notion of addition for star bodies is radial addition. Here a set $L \subset \mathbb{R}^n$ is a star body, if it is sharshaped with respect to the origin and has a continuous *radial function* $\rho(L,\cdot)$, which is defined for $u \in S^{n-1}$ by

$$\rho(L,u) = \max\{t \geq 0 : t\,u \in L\}.$$

Note that $\rho(L,\cdot)$ on S^{n-1} determines L. Let $\mathcal{S}^n$ denote the set of star bodies in $\mathbb{R}^n$. Then the radial sum $L_1 \tilde{+} L_2$ of $L_1, L_2 \in \mathcal{S}^n$ is given by

$$\rho(L_1 \tilde{+} L_2, \cdot) = \rho(L_1,\cdot) + \rho(L_2,\cdot)$$

and $L_1 \tilde{+} L_2 \in \mathcal{S}^n$. Radial addition and volume are the fundamental notions in the dual Brunn-Minkowski theory (see [16]). Here we consider star body valued functional on $\mathcal{K}_0^n$ that are valuations with respect to radial addition. Note that the trivial operators, $K \mapsto c_0\,K \tilde{+} c_1(-K)$, $c_0, c_1 \geq 0$, are $\mathrm{GL}(n)$ covariant and valuations with respect to radial addition.

Theorem 12 ([49]). *An operator* $\mathrm{Z} : \mathcal{P}_0^n \to \mathcal{S}^n$ *is a non-trivial* $\mathrm{GL}(n)$ *covariant valuation if and only if there is a constant* $c \geq 0$ *such that*

$$\mathrm{Z}\,P = c\,\mathrm{I}\,P^*$$

for every $P \in \mathcal{P}_0^n$.

Here $\mathrm{I}\,P^*$ is the *intersection body* of $P^* \in \mathcal{P}_0^n$, that is, the star body whose radial function is given by

$$\rho(\mathrm{I}\,P^*, u) = \mathrm{vol}(P^* \cap u^\perp) \quad \text{for } u \in S^{n-1},$$

where $P^* \cap u^\perp$ denotes the intersection of P^* with the subspace orthogonal to u. Intersections bodies first appear in Busemann's [8] theory of area in Finsler spaces and they were first explicitly defined and named by Lutwak [54]. Intersection bodies turned out to be critical for the solution of the *Busemann-Petty problem*: If the central hyperplane sections of an origin-symmetric convex body in $\mathbb{R}^n$ are always smaller in volume than those of another such body, is its volume also smaller? Lutwak [54] showed that the answer to the Busemann-Petty problem is affirmative if the body with the smaller sections is an intersection body of a star body. This led to the final solution that the answer is affirmative if $n \leq 4$ and negative otherwise (see [14,15,18,36,37,95,97]). Further applications of intersection bodies can be found in [9,21,22,30,38,73].

References

1. S. Alesker, *Continuous rotation invariant valuations on convex sets*, Ann. of Math. (2) **149** (1999), 977–1005.
2. S. Alesker, *Description of continuous isometry covariant valuations on convex sets*, Geom. Dedicata **74** (1999), 241–248.
3. S. Alesker, *Description of translation invariant valuations on convex sets with solution of P. McMullen's conjecture*, Geom. Funct. Anal. **11** (2001), 244–272.
4. S. Alesker, *Hard Lefschetz theorem for valuations, complex integral geometry, and unitarily invariant valuations*, J. Differential Geom. **63** (2003), 63–95.
5. S. Alesker, *Hard Lefschetz theorem for valuations and related questions of integral geometry*, Geometric aspects of functional analysis, Lecture Notes in Math., vol. 1850, Springer, Berlin, 2004, 9–20.
6. S. Alesker, *The multiplicative structure on continuous polynomial valuations*, Geom. Funct. Anal. **14** (2004), 1–26.
7. S. Alesker, SU(2)*-invariant valuations*, Geometric aspects of functional analysis, Lecture Notes in Math., vol. 1850, Springer, Berlin, 2004, 21–29.
8. H. Busemann, *A theorem on convex bodies of the Brunn-Minkowski type*, Proc. Nat. Acad. Sci. U. S. A. **35** (1949), 27–31.
9. H. Busemann, *Volume in terms of concurrent cross-sections*, Pacific J. Math. **3** (1953), 1–12.
10. E. Calabi, P. Olver, and A. Tannenbaum, *Affine geometry, curve flows, and invariant numerical approximation*, Adv. Math. **124** (1996), 154–196.
11. S. Campi and P. Gronchi, *The* L^p*-Busemann-Petty centroid inequality*, Adv. Math. **167** (2002), 128–141.
12. B. Chen, *A simplified elementary proof of Hadwiger's volume theorem*, Geom. Dedicata **105** (2004), 107–120.
13. P. Dulio and C. Peri, *Invariant valuations on spherical star sets*, Rend. Circ. Mat. Palermo (2) Suppl. (2000), no. 65, part II, 81–92.
14. R. Gardner, *Intersection bodies and the Busemann-Petty problem*, Trans. Amer. Math. Soc. **342** (1994), 435–445.
15. R. Gardner, *A positive answer to the Busemann-Petty problem in three dimensions*, Ann. of Math. (2) **140** (1994), 435–447.
16. R. Gardner, *Geometric tomography*, Cambridge University Press, Cambridge, 1995.
17. R. Gardner and A. Giannopoulos, *p-cross-section bodies*, Indiana Univ. Math. J. **48** (1999), 593–613.
18. R. Gardner, A. Koldobsky, and T. Schlumprecht, *An analytic solution to the Busemann-Petty problem on sections of convex bodies*, Ann. of Math. (2) **149** (1999), 691–703.
19. J. Gates, D. Hug, and R. Schneider, *Valuations on convex sets of oriented hyperplanes*, Discrete Comput. Geom. **33** (2005), 57–65.
20. A. Giannopoulos and M. Papadimitrakis, *Isotropic surface area measures*, Mathematika **46** (1999), 1–13.
21. P. Goodey, E. Lutwak, and W. Weil, *Functional analytic characterizations*

of classes of convex bodies, Math. Z. **222** (1996), 363–381.
22. E. Grinberg and G. Zhang, *Convolutions, transforms, and convex bodies*, Proc. London Math. Soc. (3) **78** (1999), 77–115.
23. P. M. Gruber, *Asymptotic estimates for best and stepwise approximation of convex bodies I*, Forum Math. **5** (1993), 281–297.
24. P. M. Gruber, *Asymptotic estimates for best and stepwise approximation of convex bodies II*, Forum Math. **5** (1993), 521–538.
25. O. Guleryuz, E. Lutwak, D. Yang, and G. Zhang, *Information-theoretic inequalities for contoured probability distributions*, IEEE Trans. Inform. Theory **48** (2002), 2377–2383.
26. H. Hadwiger, *Vorlesungen über Inhalt, Oberfläche und Isoperimetrie*, Springer, Berlin, 1957.
27. H. Hadwiger and R. Schneider, *Vektorielle Integralgeometrie*, Elem. Math. **26** (1971), 49–57.
28. D. Hug, *Contributions to affine surface area*, Manuscripta Math. **91** (1996), 283–301.
29. D. Hug, *Curvature relations and affine surface area for a general convex body and its polar*, Result. Math. **29** (1996), 233–248.
30. N. J. Kalton and A. Koldobsky, *Intersection bodies and L_p spaces*, Adv. Math., to appear.
31. M. Kiderlen, *Blaschke and Minkowksi endomorphisms of convex bodies*, Trans. Amer. Math. Soc., to appear.
32. D. A. Klain, *A short proof of Hadwiger's characterization theorem*, Mathematika **42** (1995), 329–339.
33. D. A. Klain, *Star valuations and dual mixed volumes*, Adv. Math. **121** (1996), 80–101.
34. D. A. Klain, *Invariant valuations on star-shaped sets*, Adv. Math. **125** (1997), 95–113.
35. D. A. Klain and G.-C. Rota, *Introduction to geometric probability*, Cambridge University Press, Cambridge, 1997.
36. A. Koldobsky, *Intersection bodies in $\mathbf{R}^4$*, Adv. Math. **136** (1998), 1–14.
37. A. Koldobsky, *Intersection bodies, positive definite distributions, and the Busemann-Petty problem*, Amer. J. Math. **120** (1998), 827–840.
38. A. Koldobsky, *A functional analytic approach to intersection bodies*, Geom. Funct. Anal. **10** (2000), 1507–1526.
39. K. Leichtweiß, *Affine geometry of convex bodies*, Johann Ambrosius Barth, Heidelberg, 1998.
40. J. Lindenstrauss and V. D. Milman, *The local theory of normed spaces and its applications to convexity*, Handbook of convex geometry, Vol. A, B, North-Holland, Amsterdam, 1993, 1149–1220.
41. M. Ludwig, *Asymptotic approximation of smooth convex bodies by general polytopes*, Mathematika **46** (1999), 103–125.
42. M. Ludwig, *A characterization of affine length and asymptotic approximation of convex discs*, Abh. Math. Semin. Univ. Hamb. **69** (1999), 75–88.
43. M. Ludwig, *Upper semicontinuous valuations on the space of convex discs*, Geom. Dedicata **80** (2000), 263–279.

44. M. Ludwig, *Moment vectors of polytopes*, Rend. Circ. Mat. Palermo (2) Suppl. (2002), no. 70, part II, 123–138.
45. M. Ludwig, *Projection bodies and valuations*, Adv. Math. **172** (2002), 158–168.
46. M. Ludwig, *Valuations on polytopes containing the origin in their interiors*, Adv. Math. **170** (2002), 239–256.
47. M. Ludwig, *Ellipsoids and matrix valued valuations*, Duke Math. J. **119** (2003), 159–188.
48. M. Ludwig, *Minkowski valuations*, Trans. Amer. Math. Soc. **357** (2005), 4157–4190.
49. M. Ludwig, *Intersection bodies and valuations*, preprint.
50. M. Ludwig, *Convex body valued valuations*, in preparation.
51. M. Ludwig and M. Reitzner, *A characterization of affine surface area*, Adv. Math. **147** (1999), 138–172.
52. M. Ludwig and M. Reitzner, *A classification of $SL(n)$ invariant valuations*, in preparation.
53. E. Lutwak, *On some affine isoperimetric inequalities*, J. Differential Geom. **23** (1986), 1–13.
54. E. Lutwak, *Intersection bodies and dual mixed volumes*, Adv. Math. **71** (1988), 232–261.
55. E. Lutwak, *Centroid bodies and dual mixed volumes*, Proc. Lond. Math. Soc. **60** (1990), 365–391.
56. E. Lutwak, *Extended affine surface area*, Adv. Math. **85** (1991), 39–68.
57. E. Lutwak, *The Brunn-Minkowski-Firey theory. I. Mixed volumes and the Minkowski problem*, J. Differential Geom. **38** (1993), 131–150.
58. E. Lutwak, *Inequalities for mixed projection bodies*, Trans. Amer. Math. Soc. **339** (1993), 901–916.
59. E. Lutwak, *Selected affine isoperimetric inequalities*, Handbook of Convex Geometry, Vol. A (P.M. Gruber and J.M. Wills, eds.), North-Holland, Amsterdam, 1993, 151–176.
60. E. Lutwak, *The Brunn-Minkowski-Firey theory. II: Affine and geominimal surface areas*, Adv. Math. **118** (1996), 244–294.
61. E. Lutwak and V. Oliker, *On the regularity of solutions to a generalization of the Minkowski problem*, J. Differential Geom. **41** (1995), 227–246.
62. E. Lutwak, D. Yang, and G. Zhang, *L_p affine isoperimetric inequalities*, J. Differential Geom. **56** (2000), 111–132.
63. E. Lutwak, D. Yang, and G. Zhang, *A new ellipsoid associated with convex bodies*, Duke Math. J. **104** (2000), 375–390.
64. E. Lutwak, D. Yang, and G. Zhang, *A new affine invariant for polytopes and Schneider's projection problem*, Trans. Amer. Math. Soc. **353** (2001), 1767–1779.
65. E. Lutwak, D. Yang, and G. Zhang, *The Cramer-Rao inequality for star bodies*, Duke Math. J. **112** (2002), 59–81.
66. E. Lutwak and G. Zhang, *Blaschke-Santaló inequalities*, J. Differential Geom. **47** (1997), 1–16.
67. P. McMullen, *Valuations and dissections*, Handbook of Convex Geometry,

Vol. B (P.M. Gruber and J.M. Wills, eds.), North-Holland, Amsterdam, 1993, 933–990.
68. P. McMullen, *Isometry covariant valuations on convex bodies*, Rend. Circ. Mat. Palermo (2) Suppl. (1997), no. 50, 259–271.
69. P. McMullen, *Polytope algebras, tensor weights and piecewise polynomials*, Intuitive geometry (Budapest, 1995), János Bolyai Math. Soc., Budapest, 1997, 159–177.
70. P. McMullen and R. Schneider, *Valuations on convex bodies*, Convexity and its applications (P.M. Gruber and J.M. Wills, eds.), Birkhäuser, 1983, 170–247.
71. M. Meyer and E. Werner, *The Santalo-regions of a convex body*, Trans. Am. Math. Soc. **350** (1998), 4569–4591.
72. M. Meyer and E. Werner, *On the p-affine surface area*, Adv. Math. **152** (2000), 288–313.
73. V. D. Milman and A. Pajor, *Isotropic position and inertia ellipsoids and zonoids of the unit ball of a normed n-dimensional space*, Geometric aspects of functional analysis (1987–88), Lecture Notes in Math., vol. 1376, Springer, Berlin, 1989, 64–104.
74. C. M. Petty, *Centroid surfaces*, Pacific J. Math. **11** (1961), 1535–1547.
75. C. M. Petty, *Projection bodies*, Proc. Colloquium on Convexity (Copenhagen, 1965), Kobenhavns Univ. Mat. Inst., Copenhagen, 1967, 234–241.
76. C. M. Petty, *Isoperimetric problems*, Proceedings of the Conference on Convexity and Combinatorial Geometry (Univ. Oklahoma, Norman, Okla., 1971), Dept. Math., Univ. Oklahoma, Norman, Okla., 1971, 26–41.
77. C. A. Rogers and G. C. Shephard, *The difference body of a convex body*, Arch. Math. **8** (1957), 220–233.
78. R. Schneider, *Krümmungsschwerpunkte konvexer Körper. II*, Abh. Math. Sem. Univ. Hamburg **37** (1972), 204–217.
79. R. Schneider, *Equivariant endomorphisms of the space of convex bodies*, Trans. Amer. Math. Soc. **194** (1974), 53–78.
80. R. Schneider, *Affine-invariant approximation by convex polytopes*, Stud. Sci. Math. Hung. **21** (1986), 401–408.
81. R. Schneider, *Convex bodies: the Brunn–Minkowski theory*, Cambridge Univ. Press, Cambridge, 1993.
82. R. Schneider, *Simple valuations on convex bodies*, Mathematika **43** (1996), 32–39.
83. R. Schneider, *Tensor valuations on convex bodies and integral geometry*, Rend. Circ. Mat. Palermo (2) Suppl. (2000) no. 65, part I , 295–316.
84. R. Schneider and R. Schuster, *Tensor valuations on convex bodies and integral geometry. II*, Rend. Circ. Mat. Palermo (2) Suppl. (2002), no. 70, part II, 295–314.
85. F.E. Schuster, *Convolutions and multiplier transformations of convex bodies*, preprint.
86. C. Schütt, *Random polytopes and affine surface area*, Math. Nachr. **170** (1994), 227–249.
87. C. Schütt and E. Werner, *Surface bodies and p-affine surface area*, Adv.

Math. **187** (2004), 98–145.
88. A. C. Thompson, *Minkowski geometry*, Cambridge University Press, Cambridge, 1996.
89. N. S. Trudinger and X.-J. Wang, *The Bernstein problem for affine maximal hypersurfaces*, Invent. Math. **140** (2000), 399–422.
90. N. S. Trudinger and X.-J. Wang, *The affine Plateau problem*, J. Amer. Math. Soc. **18** (2005), 253–289.
91. X.-J. Wang, *Affine maximal hypersurfaces*, Proceedings of the International Congress of Mathematicians, Vol. III (Beijing, 2002) (Beijing), Higher Ed. Press, 2002, 221–231.
92. E. Werner, *Illumination bodies and affine surface area*, Stud. Math. **110** (1994), 257–269.
93. E. Werner, *A general geometric construction for affine surface area*, Studia Math. **132** (1999), 227–238.
94. G. Zhang, *Restricted chord projection and affine inequalities*, Geom. Dedicata **39** (1991), 213–222.
95. G. Zhang, *Centered bodies and dual mixed volumes*, Trans. Am. Math. Soc. **345** (1994), 777–801.
96. G. Zhang, *The affine Sobolev inequality*, J. Differential Geom. **53** (1999), 183–202.
97. G. Zhang, *A positive solution to the Busemann-Petty problem in* $\mathbf{R}^4$, Ann. of Math. (2) **149** (1999), 535–543.

CROFTON MEASURES IN PROJECTIVE FINSLER SPACES

ROLF SCHNEIDER*

Mathematisches Institut,
Albert-Ludwigs-Universität,
Eckerstr. 1, D-79104, Freiburg i.Br., Germany
E-mail: rolf.schneider@math.uni-freiburg.de

The classical Crofton formula of integral geometry expresses the area of a k-dimensional surface in Euclidean space as an integral, with respect to an invariant measure, of the number of intersection points with affine flats of the complementary dimension. This paper surveys attempts that have been made to obtain similar results in finite-dimensional normed spaces and in projective Finsler spaces. The stress is on relations to the theory of general (non-smooth) convex bodies and in particular to the geometry of zonoids.

The starting point of this introductory survey is a classical formula of integral geometry in Euclidean space. It interprets the volume of a submanifold as the measure of the set of flats of complementary dimension that hit the submanifold (counted with multiplicities). More precisely, let M be a k-dimensional C^1 submanifold of Euclidean space $\mathbb{R}^n$, where $n \geq 2$ and $k \in \{1, \dots, n-1\}$, and let λ_k be the k-dimensional differential-geometric surface area measure. Let $A(n,j)$ denote the affine Grassmannian of j-flats (j-dimensional affine subspaces) of $\mathbb{R}^n$. It carries an essentially unique Haar measure μ_j (a rigid motion invariant positive Borel measure which is finite on compact sets and not identically zero). An integral-geometric result known as the **Crofton formula** says that

$$\int_{A(n,n-k)} \operatorname{card}(E \cap M)\, \mu_{n-k}(\mathrm{d}E) = a_{nk}\lambda_k(M), \tag{1}$$

where the constant a_{nk} depends on the normalization of the measure μ_{n-k} (see, e.g., Santaló[37], p. 245, (14.69)).

*Work partially supported by the European Network PHD, FP6 Marie Curie Actions, RTN, Contract MCRN-511953.

In the following, we are interested in generalizations of (1) beyond Euclidean geometry and in reverse questions of the following kind. Suppose we are given a notion of area of k-dimensional surfaces that replaces λ_k; is it possible to represent it in the form (1)? In other words, we ask whether there exists a measure replacing μ_{n-k} in (1) so that the analogue of (1) holds for a large class of submanifolds M. By a **measure** on a locally compact space we understand in this survey a signed measure on the Borel sets of the space, which is finite on compact sets. A **positive measure** is a measure attaining only nonnegative values. A measure satisfying the generalized version of (1) as explained will be called a **Crofton measure** for the given notion of area. If such a formula exists, it connects metric notions, namely areas, with affine notions, namely flats; therefore, the existence can only be expected in situations where metric and affine structures are tied together in some way. A natural geometric environment of this kind is provided by the (general) projective Finsler spaces. Projective Finsler metrics are a special case of the projective metrics appearing in Hilbert's fourth problem. It is, in fact, the integral-geometric approach to Hilbert's fourth problem from which a natural development has led to the investigation of Crofton type formulas in projective Finsler spaces. We begin, therefore, our survey with a brief sketch of Hilbert's fourth problem and the role of integral geometry in its treatment.

1. The Integral-geometric Approach to Hilbert's Fourth Problem

The fourth problem in Hilbert's famous collection of 1900, entitled 'Problem von der Geraden als kürzester Verbindung zweier Punkte' (Problem of the straight line as the shortest connection of two points), was originally motivated by Hilbert's investigations into the foundations of geometry. Roughly speaking, it asks for the geometries, defined axiomatically, in which there exists a notion of length for which line segments are the shortest connections of their endpoints. The problem has later seen many transformations, generalizations as well as specializations, and we formulate here only a special case in later terminology:

H4. *Given an open convex subset C of $\mathbb{R}^n$, determine all complete projective metrics on C.*

A metric d on C is called **projective** if it is continuous and satisfies

$$d(p,q) + d(q,r) = d(p,r) \tag{2}$$

whenever p, q, r are points on a line, in this order. A metric satisfying (2) is also called **linearly additive**. For a given metric d, the length of a continuous parameterized curve $\gamma : [a, b] \to C$ is defined by

$$L(\gamma) := \sup \sum_{i=1}^{k} d(\gamma(t_{i-1}), \gamma(t_i)),$$

where the supremum is taken over all subdivisions $a = t_0 < t_1 < \cdots < t_k = b$, $k \in \mathbb{N}$. For the segment $\overline{pq}$ with endpoints p and q, (2) implies $L(\overline{pq}) = d(p, q)$, and every continuous curve γ with endpoints p and q satisfies $L(\gamma) \geq L(\overline{pq})$. Conversely, let L be a notion of curve length for which $L(\overline{pq}) \leq L(\gamma)$ holds for every continuous curve from p to q. It induces a metric d by $d(p, q) := \inf L(\gamma)$, where the infimum is taken over all continuous curves from p to q. This metric d then satisfies $d(p, q) = L(\overline{pq})$ and, hence, also (2).

The notion of a projective metric is natural and fundamental: two basic structures, a metric and the linear structure of an affine space, are tied together by the compatibility condition (2). The determination of all projective metrics, however, is not an easy task. It is interesting to quote here from Busemann[21]: "... Specifically, Hilbert asks for the construction of all these metrics and the study of the individual geometries. It is clear from Hilbert's comments that he was not aware of the immense number of these metrics, so that the second part of the problem is not a well posed question and has inevitably been replaced by the investigation of special, or special classes of, interesting geometries."

There are two classical examples of projective metrics, already given by Hilbert with the formulation of his fourth problem. The first example is that of a **Minkowski space**, that is, $\mathbb{R}^n$ with the metric induced by a norm $||\cdot||$. In that case, the distance defined by $d(x, y) = ||x - y||$ is invariant under translations. The metrics coming from a norm are precisely the translation invariant projective metrics on $\mathbb{R}^n$. The second example is what is now called a **Hilbert geometry**. Here it is assumed that the open convex set C is bounded. For $x, y \in C$, $x \neq y$, let a, b be the points where the line through x and y meets the boundary of C, so that a, x, y, b appear in this order on the line. With an auxiliary Euclidean norm $|\cdot|$, define

$$d(x, y) := \ln \frac{|x - b||y - a|}{|y - b||x - a|}. \tag{3}$$

Then d is a projective metric on C.

Further examples of projective metrics do not easily come to mind. However, a wealth of them can be constructed by a nice integral-geometric approach suggested by Busemann, around 1960 (see Busemann[20,19]). Take any positive measure μ on the space $A(n, n-1)$ of hyperplanes of $\mathbb{R}^n$ which satisfies

$$\mu(\{H \in A(n, n-1) : p \in H\}) = 0 \qquad \text{for each } p \in \mathbb{R}^n \tag{4}$$

and

$$0 < \mu(\{H \in A(n, n-1) : H \cap \overline{qp} \neq \emptyset\}) < \infty \qquad \text{for } p \neq q. \tag{5}$$

If we define

$$d(p,q) := \mu(\{H \in A(n, n-1) : H \cap \overline{pq} \neq \emptyset\}) \tag{6}$$

for $p, q \in \mathbb{R}^n$, then d is a projective metric. The triangle inequality $d(p,r) \leq d(p,q) + d(q,r)$ follows from the fact that every hyperplane meeting the segment $\overline{pq}$ has to meet one of the segments $\overline{pq}$ or $\overline{qr}$. When q lies in the segment $\overline{pr}$, then obviously (2) holds.

The question immediately arises whether this construction produces all projective metrics on $\mathbb{R}^n$. For the plane, the answer is affirmative. In independent work of Pogorelov[36], Ambartzumian[13], Alexander[1], the following result was shown, in varying degrees of generality and with different formulations:

Theorem 1.1. *The equation* (6) *establishes a linear isomorphism between the cone of projective metrics on* $\mathbb{R}^2$ *and the cone of positive measures* μ *on the space* $A(2,1)$ *of lines in* $\mathbb{R}^2$ *satisfying* (4) *and* (5).

In dimensions greater than two, the situation is different; not every projective metric on $\mathbb{R}^n$, $n \geq 3$, can be obtained by Busemann's construction with a measure. This is already seen from the example of Minkowski spaces. The study of this case leads us in the next section to the zonoid equation, which plays an important role later on.

2. The Zonoid Equation

It is convenient in the following to use an auxiliary Euclidean structure on $\mathbb{R}^n$, given by a scalar product $\langle \cdot, \cdot \rangle$ with induced Euclidean norm $|\cdot|$. Then $S^{n-1} := \{u \in \mathbb{R}^n : |u| = 1\}$ is the Euclidean unit sphere. By σ we denote the spherical Lebesgue measure on S^{n-1}. With the aid of the scalar product, we parameterize the hyperplanes $H \in A(n, n-1)$ in the form

$$H_{u,t} = \{x \in \mathbb{R}^n : \langle x, u \rangle = t\}$$

with $u \in S^{n-1}$ and $t \in \mathbb{R}$.

Now let $||\cdot||$ be a norm on $\mathbb{R}^n$. Then

$$B := \{x \in \mathbb{R}^n : ||x|| \leq 1\}$$

is the unit ball of the Minkowski space $(\mathbb{R}^n, ||\cdot||)$. We use the scalar product to identify $\mathbb{R}^n$ with its dual space, and hence identify the dual unit ball with the polar body of B,

$$B^o := \{u \in \mathbb{R}^n : \langle u, x\rangle \leq 1 \ \forall x \in B\}.$$

Then

$$h(B^o, \cdot) = ||\cdot||$$

is the support function of B^o.

Suppose now that the projective metric d with $d(x,y) = ||x-y||$ can be generated by (6), with a measure μ on $A(n, n-1)$ satisfying (4) and (5). It can be shown (Alexander[2]) that such a measure is unique. Since d is translation invariant, the uniqueness implies that μ is translation invariant. We call the translation invariant measure μ on $A(n, n-1)$ satisfying

$$\|x-y\| = \int_{A(n,n-1)} \operatorname{card}(H \cap \overline{xy})\, \mu(\mathrm{d}H) \qquad \text{for } x, y \in \mathbb{R}^n \tag{7}$$

a **Crofton measure** for the norm $\|\cdot\|$. This is the simplest case of a Crofton measure for which there is a non-trivial existence problem; more general Crofton measures are the main topic of this survey.

The translation invariant (and, as generally assumed, locally finite) measure μ can be decomposed: there exists a finite measure φ on the sphere S^{n-1} such that

$$\int_{A(n,n-1)} f \,\mathrm{d}\mu = \int_{S^{n-1}} \int_{-\infty}^{\infty} f(H_{u,t}) \,\mathrm{d}t \,\varphi(\mathrm{d}u) \tag{8}$$

holds for every nonnegative measurable function f (see, for example, Schneider and Weil[47], Satz 4.1.1, also for measures on $A(n,j)$). Since $H_{u,t} = H_{-u,-t}$, the measure φ can be assumed to be even (i.e., to satisfy $\varphi(A) = \varphi(-A)$ for Borel sets $A \subset S^{n-1}$). Now the assumed representation (6), or (7), gives

$$||x|| = d(0,x) = \int_{S^{n-1}} \int_{-\infty}^{\infty} \mathbf{1}\{H_{u,t} \cap \overline{0x} \neq \emptyset\} \,\mathrm{d}t \,\varphi(\mathrm{d}u),$$

hence

$$h(B^o, x) = \int_{S^{n-1}} |\langle u, x\rangle| \,\varphi(\mathrm{d}u). \tag{9}$$

The equation

$$f(x) = \int_{S^{n-1}} |\langle u, x \rangle| \, \varphi(\mathrm{d}u), \qquad x \in \mathbb{R}^n, \tag{10}$$

or specialized to

$$f(x) = \int_{S^{n-1}} |\langle u, x \rangle| g(u) \, \sigma(\mathrm{d}u), \qquad x \in \mathbb{R}^n, \tag{11}$$

where f is a given even function and either the measure φ or the measurable function g is required, is known as the **zonoid equation**. This terminology has the following geometric background. Let $Z \subset \mathbb{R}^n$ be a convex body whose support function can be represented in the form

$$h(Z, v) = \int_{S^{n-1}} |\langle u, v \rangle| \, \varphi(\mathrm{d}u), \qquad v \in S^{n-1}, \tag{12}$$

with a positive even measure φ on S^{n-1}. If φ has finite support, then $h(Z, \cdot)$ is the sum of the support functions of finitely many (line) segments, hence Z is the vector sum of finitely many segments. The convex bodies which are sums of finitely many segments are precisely the polytopes all of whose faces are centrally symmetric; they are called **zonotopes**, since their facets are arranged in zones. A convex body which can be approximated, in the Hausdorff metric, by zonotopes, is called a **zonoid**. Every zonoid has a centre of symmetry. A convex body with symmetry centre 0 is called **centred**. The convex bodies Z representable by (12) with a positive measure φ are precisely the centred zonoids. If Z, after a suitable translation, can be represented by (12) with a measure φ (not necessarily positive), then it is called a **generalized zonoid**. If (12) holds with an even measure φ, then φ is called the **generating measure** of Z, and a continuous density of φ with respect to σ, if it exists, is called the **generating function** of Z.

If (10) is satisfied with an even measure φ, then we can use (8) to define a translation invariant measure μ on $A(n, n-1)$, and this satisfies (7).

About the zonoid equation, the following is known, for example. An even measure φ satisfying (10) for given f is uniquely determined. If the even function f is sufficiently often differentiable (where 'sufficient' increases with the dimension), then (11) has a continuous solution g. This result goes back to Blaschke; a rigorous proof specifying differentiability assumptions was given in Schneider[38]; or see Schneider[40], Th. 3.5.3. From this it follows that in the set of centrally symmetric convex bodies the generalized zonoids are dense, whereas the set of zonoids is closed and nowhere dense. Not every centrally symmetric convex body is a generalized zonoid.

It follows from Schneider[39], Lemma 6.1, that a convex body which has a face without a centre of symmetry cannot be a generalized zonoid.

For the existence of Crofton measures for norms, this has the following consequences.

Theorem 2.1. *A Crofton measure for a norm exists if and only if the polar unit ball of the norm is a generalized zonoid.*

For every sufficiently smooth norm on $\mathbb{R}^n$ there exists a Crofton measure, even one with a continuous density with respect to the motion invariant measure.

There are norms, for example that of ℓ_∞^n, for which no Crofton measure exists.

The existence of a positive Crofton measure is a much more restrictive property of a norm. To see this, choose m points $p_1, \dots, p_m \in \mathbb{R}^n$ and integers $N_1, \dots, N_m$ with $\sum_{i=1}^m N_i = 1$. Let H be a hyperplane not incident with one of the points $p_1, \dots, p_m$, and let H^+, H^- be the two closed halfspaces bounded by H. Then

$$\sum_{i<j} \mathbf{1}\{\overline{p_i p_j} \cap H \neq \emptyset\} N_i N_j$$

$$= \left(\sum_{p_i \in H^+} N_i \right) \left(\sum_{p_j \in H^-} N_j \right) = \left(\sum_{p_i \in H^+} N_i \right) \left(1 - \sum_{p_i \in H^+} N_i \right) \leq 0.$$

If now (7) holds with a positive measure μ, then integration over all hyperplanes with respect to μ gives

$$\sum_{i<j} d(p_i, p_j) N_i N_j \leq 0. \tag{13}$$

A metric d satisfying (13) for all m-tuples $(N_1, \dots, N_m)$ of integers which sum to one, all $m \in \mathbb{N}$, and all $p_1, \dots, p_m \in \mathbb{R}^n$, is called a **hypermetric**. We say that the Minkowski space $(\mathbb{R}^n, ||\cdot||)$ is **hypermetric** if its induced metric d is a hypermetric.

We have shown: if B^o is a zonoid, then $(\mathbb{R}^n, ||\cdot||)$ with $||\cdot|| = h(B^o, \cdot)$ is hypermetric. The converse is also true, but lies deeper (references are in Schneider and Weil[46], p. 301). We collect the results in the following theorem.

Theorem 2.2. *For a norm $||\cdot||$ on $\mathbb{R}^n$ with unit ball B, the following conditions are equivalent:*

(a) *There exists a positive Crofton measure for* $||\cdot||$.

(b) *The polar unit ball* B^o *is a zonoid.*

(c) *The Minkowski space* $(\mathbb{R}^n, ||\cdot||)$ *is hypermetric.*

The theorems of this section already indicate the role that zonoids and generalized zonoids play for the study of Crofton measures. This important role becomes even more evident in the generalizations considered in the sequel.

Surveys on zonoids were written by Bolker[17], Schneider and Weil[46], Goodey and Weil[31]. For background material on hypermetrics, we refer to Kelly[34] and to Deza and Laurent[24].

3. Projective Finsler Spaces

Projective Finsler spaces are intermediate between Minkowski spaces and general projective metrics, in different respects. First, for a Finsler space, every tangent space is a Minkowski space. Second, sufficiently smooth projective metrics are induced from Finsler metrics, and projective metrics can be approximated, uniformly on compact sets, by smooth projective Finsler metrics.

In the following, we canonically identify the tangent space $T_x\mathbb{R}^n$ of $\mathbb{R}^n$ at the point x with $\mathbb{R}^n$. Projective Finsler metrics can be defined on open convex subsets C of $\mathbb{R}^n$. We define a (general) **Finsler metric** on C as a continuous function $F : C \times \mathbb{R}^n \to [0, \infty)$ with the property that $F(x, \cdot)$ is a norm on $\mathbb{R}^n$, for each $x \in C$. Thus, we consider here only symmetric Finsler metrics, and we dispense with the differentiability assumptions common in Finsler geometry; this is a natural approach when we want to generalize arbitrary Minkowski spaces and Hilbert geometries. The **length** of a parameterized C^1 curve $\gamma : [a, b] \to C$ is defined by $\int_a^b F(\gamma(t), \gamma'(t))\,\mathrm{d}t$. The Finsler space (C, F) is called **projective** if line segments are shortest curves connecting their endpoints; here it is not required that shortest connections are unique. The Finsler metric F induces a metric d_F by defining $d_F(p, q)$ as the infimum of the lengths of all piecewise C^1 curves connecting the points $p, q \in C$. If (C, F) is projective, then the segment $\overline{pq}$ has length $d_F(p, q)$. If the metric d_F is a hypermetric, then the Finsler space (C, F) is called **hypermetric**.

We have already seen in Theorem 2.1 that, with regard to the existence of Crofton measures, there are essential differences between the smooth and non-smooth cases. For this reason, we have also to consider smooth Finsler metrics. The Finsler metric F is said to be **smooth** if F is of class C^∞ on $C \times (\mathbb{R}^n \setminus \{0\})$. (The further assumption common in Finsler geometry, namely that $F(x,\cdot)^2$ has positive definite Hessian on $\mathbb{R}^n \setminus \{0\}$, is not needed in the following.) If F is a (smooth) Finsler metric on C, then (C, F) is called a (smooth) **Finsler space**.

Let (C, F) be a Finsler space. For each $x \in C$, $\|\cdot\|_x$ is a norm on $T_x\mathbb{R}^n = \mathbb{R}^n$, and the sets

$$B_x := \{\xi \in \mathbb{R}^n : F(x,\xi) \leq 1\}, \qquad B_x^o := \{u \in \mathbb{R}^n : \langle u, \xi \rangle \leq 1 \ \forall \xi \in B_x\}$$

are, respectively, the unit ball and polar unit ball of this norm. The body B_x is also called the **indicatrix**, and B_x^o is called the **figuratrix**, of the Finsler metric F at x. The role of the figuratrix in the calculus of variations is explained in Blaschke[16].

If the Finsler space is hypermetric, then it can be shown (see Alexander[2]) that also the metric induced by $\|\cdot\|_x$ is a hypermetric, hence B_x^o is a zonoid, for each $x \in C$.

Examples for projective Finsler spaces are, of course, the Minkowski spaces, but also the Hilbert geometries, since their metric is projective and is induced from a Finsler metric. This is seen as follows. Let $K \subset \mathbb{R}^n$ be a convex body with interior points and let C be its interior. For distinct points $x, y \in C$, the Hilbert distance $d(x, y)$ is defined by (3); together with $d(x, x) = 0$ this defines a projective metric d on C for which (C, d) is complete. With x, y, a, b as in (3), let $u \in \mathbb{R}^n \setminus \{0\}$ be a vector such that $b = x + t_1 u$, $a = x - t_2 u$ with $t_1, t_2 > 0$. For $\lambda > 0$ with $x + \lambda u \in C$ we have

$$d(x, x + \lambda u) = \ln \frac{t_1(t_2 + \lambda)}{(t_1 - \lambda)t_2}$$

and hence

$$F(x, u) := \lim_{\lambda \to 0} \frac{1}{\lambda} d(x, x + \lambda u) = \frac{1}{t_1} + \frac{1}{t_2}.$$

This gives

$$\begin{aligned} \int_0^1 F(x + \tau(y - x), y - x)\, \mathrm{d}\tau &= \int_0^1 \left(\frac{1}{t_1 - \tau} + \frac{1}{t_2 + \tau} \right) \mathrm{d}\tau \\ &= \ln \frac{t_1(t_2 + 1)}{(t_1 - 1)t_2} = d(x, y). \end{aligned}$$

Denoting by $\rho(M, \cdot)$ the radial function of a convex body M, we have

$$t_1 = \rho(K - x, u), \qquad t_2 = \rho(K - x, -u)$$

and hence (using Schneider[40], Remark 1.7.7)

$$F(x,u) = h((K-x)^o, u) + h((K-x)^o, -u) = h((K-x)^o - (K-x)^o, u).$$

We put $(K-x)^o =: K^x$ and have $K^x - K^x = \mathrm{D}K^x$, where D denotes the difference body operator, thus

$$F(x,\cdot) = h(\mathrm{D}K^x, \cdot)$$

(defining $F(x,0) := 0$). This shows that $F(x,\cdot)$ is a norm on $\mathbb{R}^n$, and its dual unit ball at $x \in C$ is given by

$$B_x^o = \mathrm{D}K^x. \tag{14}$$

Together with the observations above, this yields that (C, F) is a projective Finsler space, and its induced metric d_F is equal to d (observe that $d_F(x,y)$ can also be obtained as the infimum of the Finsler lengths of all polygonal curves joining x and y).

From now on we restrict ourselves, for simplicity of presentation, to Finsler metrics on $\mathbb{R}^n$.

We turn to Crofton measures. By a **Crofton measure** for the Finsler metric F we understand a measure η on $A(n,n-1)$ with the property that

$$d_F(p,q) = \int_{A(n,n-1)} \mathrm{card}(H \cap \overline{pq})\,\eta(\mathrm{d}H) \qquad \text{for } p,q \in \mathbb{R}^n. \tag{15}$$

We will now sketch how Pogorelov[36] established the existence of Crofton measures for smooth projective Finsler metrics, using the zonoid equation.

Let F be a smooth projective Finsler metric on $\mathbb{R}^n$. Since lines are extremals of the length integral, they must satisfy the Euler-Lagrange equations. From this, Pogorelov[36], p. 63, deduces that

$$\frac{\partial^2 F(x,\xi)}{\partial x_i \partial \xi_j} = \frac{\partial^2 F(x,\xi)}{\partial x_j \partial \xi_i} \tag{16}$$

for $x = (x_1, \ldots, x_n)$, $\xi = (\xi_1, \ldots, \xi_n) \neq 0$ and $i,j = 1,\ldots,n$. (In a special case, this was already done by Hamel[33]. See Álvarez and Fernandes[10] for a short proof of a generalization.) For every fixed $x \in \mathbb{R}^n$, Pogorelov now solves the zonoid equation for the function $F(x,\cdot)$. Since this function is of class C^∞, there exists a continuous even function $\gamma(x,\cdot)$ on S^{n-1} such that

$$F(x,\xi) = \int_{S^{n-1}} |\langle \xi, u\rangle| \gamma(x,u)\,\sigma(\mathrm{d}u) \qquad \text{for } \xi \in \mathbb{R}^n. \tag{17}$$

Different from the approach in Pogorelov[36], this can be done by using expansions in spherical harmonics, as in Schneider[38] (see also Szabó[49]). With the aid of the estimates obtained in Schneider[38], one can then show that the function γ satisfies the differentiability assumptions, also with respect to the coordinates of x, to allow the following procedure. An argument of Pogorelov[36], p. 63, deduces from (16) and (17) that

$$\int_{S_\xi} \left(\frac{\partial \gamma(x,u)}{\partial x_i} u_j - \frac{\partial \gamma(x,u)}{\partial x_j} u_i \right) \sigma(\mathrm{d}u) = 0$$

holds for $i, j = 1, \ldots, n$ and all $\xi \in S^{n-1}$, where $S_\xi := \{u \in S^{n-1} : \langle \xi, u \rangle > 0\}$. Since $\gamma(x, \cdot)$ is even, the integrand here is an odd function. An odd continuous function on S^{n-1} with vanishing integrals over all hemispheres is identically zero (a result due to Funk[27]; see Schneider[39], Korollar 3.2, for a proof of an extension, and Groemer[32], Section 3.4, for a systematic treatment of the 'hemispherical transformation' and similar transformations by means of spherical harmonics). It results that

$$\frac{\partial \gamma(x,u)}{\partial x_i} u_j - \frac{\partial \gamma(x,u)}{\partial x_j} u_i = 0$$

for $(x, u) \in \mathbb{R}^n \times S^{n-1}$. This means that, for fixed $u \in S^{n-1}$, the gradient of $\gamma(\cdot, u)$ is proportional to u, hence $\gamma(\cdot, u)$ is constant on every hyperplane with normal vector u. Therefore, we can define a continuous function $\Gamma : A(n, n-1) \to \mathbb{R}$ by letting $\Gamma(H)$, for $H \in A(n, n-1)$, be the value that $\gamma(\cdot, u)$ attains on H, if u is a unit normal vector of H (it does not matter whether we choose u or $-u$, since $\gamma(\cdot, u) = \gamma(\cdot, -u)$). Let η be the measure on $A(n, n-1)$ which has the function Γ as a density with respect to the rigid motion invariant measure (corresponding to the Euclidean structure) on $A(n, n-1)$, suitably normalized. We make this more explicit, and at the same time fix a normalization. Since $\gamma(\cdot, u)$ is constant on the hyperplane $H_{u,t}$, we can define a continuous function $g : S^{n-1} \times \mathbb{R} \to \mathbb{R}$ with $g(u, t) = g(-u, -t)$ by means of $\gamma(x, u) =: g(u, \langle x, u \rangle)$. Then we can define the measure η by

$$\int_{A(n,n-1)} f \, \mathrm{d}\eta := \int_{S^{n-1}} \int_{-\infty}^{\infty} f(H_{u,t}) g(u,t) \, \mathrm{d}t \, \sigma(\mathrm{d}u) \tag{18}$$

for every nonnegative measurable function f on $A(n, n-1)$. The signed measure η satisfies (15). This is shown by Pogorelov[36] (for $n = 2, 3$), see also Szabó[49], but will also follow from more general formulas below.

Since $F(x, \cdot) = \| \cdot \|_x$ is the support function of the polar unit ball B_x^o, we can write (17) in a form which will later be useful. So far, the following has been obtained.

Theorem 3.1. *If* $(\mathbb{R}^n, F)$ *is a smooth projective Finsler space, then there exists a uniquely determined continuous function* $g : S^{n-1} \times \mathbb{R} \to \mathbb{R}$ *with* $g(u,t) = g(-u,-t)$ *such that*

$$h(B_x^o, \xi) = \int_{S^{n-1}} |\langle \xi, u\rangle| g(u, \langle x, u\rangle)\, \sigma(\mathrm{d}u) \qquad \text{for } x, \xi \in \mathbb{R}^n. \tag{19}$$

Therefore, a Crofton measure exists for the metric d_F.

The equation (19) has, in particular, the following consequence. For $x, y \in \mathbb{R}^n$, the generating functions of B_x^o and B_y^o coincide on the subsphere s_{x-y}, where $s_u := S^{n-1} \cap u^\perp$; here $u^\perp$ is the linear subspace through 0 orthogonal to u. Thus, in a smooth projective Finsler space, the unit balls at different points are strongly tied together and can never be chosen independently. Considerations on the construction of projective Finsler metrics are found in Hamel[33] and Ambartzumian and Oganian[14].

To obtain the above results, strong smoothness assumptions are necessary, though weaker differentiability assumptions than C^∞ are sufficient. For general projective metrics, there are approximation results due to Pogorelov[36] (for $n = 2, 3$) and Szabó[49]. These authors have shown that projective metrics can be approximated suitably by smooth projective Finsler metrics. The corresponding sequences of Crofton measures will in general not contain vaguely convergent subsequences, so that the value of these approximations is limited. This is different, though, if the measures are positive. Therefore, further conclusions can be drawn in the case of projective hypermetrics. In that case, Alexander[2] has shown that also the approximating Finsler spaces are hypermetric and that, therefore, all the local polar unit balls are zonoids. The corresponding Crofton measures are then positive measures, and Alexander[2] was able to deduce the following result.

Theorem 3.2. *The equation* (6) *establishes a linear isomorphism between the cone of projective hypermetrics on* $\mathbb{R}^n$ *and the cone of positive measures* μ *on the space* $A(n, n-1)$ *of hyperplanes in* $\mathbb{R}^n$ *satisfying* (4) *and* (5).

This can be considered as a solution of Hilbert's fourth problem for the special case of projective hypermetrics. For smooth projective Finsler metrics, Pogorelov's and Szabó's results establish the existence of a Crofton measure. This is in general not positive, but has been called **quasi-positive**, since it has the following property. If $\overline{xy}$ and $\overline{yz}$ are non-collinear segments, then the measure of the set of hyperplanes intersecting both segments is positive. An effective criterion for quasi-positive measures seems to be lacking, so that the construction does not yield a complete explicit description of the smooth projective Finsler metrics. The remaining question is essentially the same as the question for the conditions which a function g in (19) has to satisfy so that the integral defines a support function. With these remarks, we leave Hilbert's fourth problem, since our main concern is the existence of Crofton measures for higher dimensional areas instead of lengths. We mention, however, that a thorough investigation of general projective metrics was undertaken by Szabó[49], and that in the smooth case there is an elegant approach via symplectic geometry due to Álvarez[5]. Álvarez[4] has also written a beautiful introduction, at an elementary level, to the planar case of Hilbert's fourth problem.

4. Notions of Area

Generalizing Hilbert's fourth problem, Busemann[20,19] proposed to study axiomatically defined notions of k-dimensional areas in n-dimensional affine spaces for which k-flats minimize area. In close connection with this, he also suggested to study the k-dimensional areas that satisfy a Crofton formula. A different approach to Crofton type results are the analytic investigations of Gelfand and Smirnov[29], Álvarez, Gelfand and Smirnov[11] on Crofton k-densities, aiming at connecting the two classical integral geometries, that of Poincaré, Blaschke, Chern on the one hand, and the integral geometry of Radon transforms on the other hand.

In the following, we are interested in Crofton type questions for areas in Minkowski and projective Finsler spaces, avoiding smoothness assumptions where possible, and emphasizing the connections to zonoid theory. General references for volumes and areas in Minkowski and Finsler spaces are the book of Thompson[50] and the article by Álvarez and Thompson[12]. The latter is a highly recommended survey article, giving a thorough introduction to volumes on normed and Finsler spaces. Insisting on the intrinsic approach as it does, it may for many readers be more satisfactory under formal aspects than the following brief ad hoc introduction, which sacrifices the pureness of approach to a more intuitive presentation and to the ease of calculations.

In an n-dimensional Finsler space, there is a natural notion of curve length, but no canonical notion of area for k-dimensional submanifolds, if $1 < k \leq n$. Instead, there are several options. This becomes already clear in the case of a Minkowski space $(\mathbb{R}^n, \|\cdot\|)$, which we consider first.

It is natural to assume, and we do this, that any area in $(\mathbb{R}^n, \|\cdot\|)$ should be invariant under the isometries of the space and thus, in particular, under translations. The n-dimensional area, or volume, is thus required to be a translation invariant measure (locally finite, as always) on $\mathbb{R}^n$ and hence is uniquely determined up to a constant factor. This factor can be chosen so that the volume of the unit ball B of the norm $\|\cdot\|$ has some desired value. In an analogous way, for $1 < k < n$, in every k-dimensional subspace a notion of k-dimensional volume is uniquely determined up to a factor. This factor, however, depends on the subspace. For that reason, there are many possibilities to define a k-dimensional area. The geometric properties of such areas can be very different, as will be seen later.

Since a k-dimensional area in a Minkowski space should be determined by its metric and thus by its unit ball, up to the freedom of choosing normalizing factors as described above, the following axioms for such an area, essentially going back to Busemann[19], are natural. Let $\mathcal{C}^k$ denote the set of all k-dimensional convex bodies in $\mathbb{R}^n$ which are centrally symmetric with respect to the origin. A **k-normalization** is a function

$$\alpha_k : \mathcal{C}^k \to \mathbb{R}^+$$

satisfying the following properties **(M1)** – **(M3)**:

(M1) α_k is invariant under linear transformations of $\mathbb{R}^n$,

(M2) α_k is continuous (with respect to the Hausdorff metric),

(M3) $\alpha_k(E_k) = \kappa_k$, if E_k is a k-dimensional ellipsoid.

If a k-normalization α_k is given, the induced **Minkowskian k-area** α_k^B on $(\mathbb{R}^n, \|\cdot\|)$ is defined by

$$\alpha_k^B(M) := \int_M \frac{\alpha_k(B \cap T_x M)}{\lambda_k(B \cap T_x M)} \lambda_k(\mathrm{d}x) \tag{20}$$

for any compact C^1 submanifold M of $\mathbb{R}^n$; here $T_x M$ is the tangent space of M at x (a subspace of $\mathbb{R}^n$, since we have identified $T_x\mathbb{R}^n$ with $\mathbb{R}^n$). We have attached the upper index B to α_k^B since the area depends on both, the k-normalization α_k and the norm $\|\cdot\|$ with unit ball B. Further, we have used the auxiliary Euclidean structure, since this is often convenient for calculations, but the definition is independent of the choice of the Euclidean metric. An intrinsic representation is given by (24) below.

In the special case $k = n-1$, a further axiom plays an important role. To formulate it, we define the **scaling function** (depending on the auxiliary

Euclidean metric) by

$$\sigma_{\alpha,B}(u) := |u| \frac{\alpha_{n-1}(B \cap u^{\perp})}{\lambda_{n-1}(B \cap u^{\perp})} \qquad \text{for } u \in \mathbb{R}^n \setminus \{0\}.$$

The fourth axiom demands:

(M4) $\sigma_{\alpha,B}$ is a norm on $\mathbb{R}^n$ (for any B).

The raison d'être for this axiom is the fact that it is equivalent to the area-minimizing property of flat regions; see Álvarez and Thompson[12], Section 4, for the explanation of several versions of this equivalence.

If condition **(M4)** is satisfied, then $\sigma_{\alpha,B}$ is the support function of a convex body $\mathbf{I}_B$. Suppose that also a Minkowskian volume is given, by an n-normalization α_n as above. Then the body $\mathbb{I}_B := (\lambda_n(B)/\alpha_n(B))\mathbf{I}_B$ is called the **isoperimetrix** of the triple $(\alpha_{n-1}, \alpha_n, B)$. The name comes from the connection with the isoperimetric problem: among all convex bodies of the same volume, precisely the homothets of the isoperimetrix have the smallest Minkowskian surface area (see Thompson[50] or Álvarez and Thompson[12]). The chosen normalization makes $\mathbb{I}_B$ independent of the auxiliary Euclidean structure.

Now let $(\mathbb{R}^n, F)$ be a Finsler space. Any Minkowskian notion of k-area, defined by a k-normalization α_k, immediately yields a notion of k-area of C^1 submanifolds M in $\mathbb{R}^n$, just by extending (20) to

$$\alpha_k^F(M) := \int_M \frac{\alpha_k(B_x \cap T_x M)}{\lambda_k(B_x \cap T_x M)} \lambda_k(\mathrm{d}x). \tag{21}$$

There are two special choices of k-normalizations α_k leading to important notions of areas, which appear particularly natural from different points of view. The first one is the trivial function

$$\alpha_k(K) := \kappa_k \qquad \text{for all } K \in \mathcal{C}^k,$$

leading to the **Busemann k-area**,

$$\beta_k(M) := \int_M \frac{\kappa_k}{\lambda_k(B_x \cap T_x M)} \lambda_k(\mathrm{d}x). \tag{22}$$

The second one is the function given by

$$\alpha_k(K) := \frac{vp(K)}{\kappa_k} \qquad \text{for all } K \in \mathcal{C}^k,$$

where $vp(K)$ is the **volume product** of K, that is, the product of the Euclidean k-dimensional volumes of K and its polar body K^o, taken in the affine hull of K; this definition is independent of the choice of the Euclidean metric. This second function yields the **Holmes–Thompson k-area,**

$$\mathrm{vol}_k(M) := \int_M \frac{\lambda_k(B_x^o|T_xM)}{\kappa_k}\,\lambda_k(\mathrm{d}x). \tag{23}$$

Here $|E$ denotes orthogonal projection to a subspace E, and a result from convex geometry was used to replace $(B_x \cap E)^o$ by $B_x^o|E$.

The definitions are also employed for $k = n$, thus

$$\beta_n(M) := \kappa_n \frac{\lambda_n(M)}{\lambda_n(B)}$$

is the Busemann volume, and

$$\mathrm{vol}_n(M) := \frac{1}{\kappa_n}\lambda_n(B^o)\lambda_n(M)$$

is the Holmes–Thompson volume of the Borel set $M \subset \mathbb{R}^n$.

The exceptional role of these two area notions is explained by the fact that they are disguised areas appearing in other contexts. The Busemann area of a k-dimensional submanifold of the Finsler space $(\mathbb{R}^n, F)$ is its k-dimensional Hausdorff measure $\mathcal{H}_F^k$ induced by the metric d_F. Proofs that the Busemann k-area of a rectifiable subset of a smooth or general Finsler space coincides with its k-dimensional Hausdorff measure $\mathcal{H}_F^k$, can be found, in different degrees of generality and with different proofs, in Busemann[18], Bellettini, Paolini and Venturini[15], Schneider[44]. The Holmes–Thompson area of a k-dimensional submanifold of a Finsler space is the symplectic (or Liouville) volume of its unit co-disc bundle with respect to the induced Finsler metric, divided by the volume of the Euclidean unit ball of dimension k.

With the help of the Hausdorff measure $\mathcal{H}_F^k$, formula (21) can be replaced by

$$\alpha_k^F(M) = \frac{1}{\kappa_k}\int_M \alpha_k(B_x \cap T_xM)\,\mathcal{H}_F^k(\mathrm{d}x), \tag{24}$$

as shown in Schneider[44]. This representation is intrinsic, that is, it no longer involves the auxiliary Euclidean metric.

In a Minkowski space (of dimension n, with unit ball B) both, the Busemann $(n-1)$-area and the Holmes–Thompson $(n-1)$-area, satisfy axiom **(M4)**. Denoting the scaling functions of the Busemann area and the Holmes–Thompson area by σ_B^{BU} and σ_B^{HT}, respectively, and the isoperimetrices (with respect to the corresponding volumes) by $\mathbb{I}_B^{\mathrm{BU}}$ and $\mathbb{I}_B^{\mathrm{HT}}$, respectively, we have

$$\sigma_B^{\mathrm{BU}}(u) = \frac{\kappa_{n-1}}{\lambda_{n-1}(B \cap u^\perp)}, \qquad u \in S^{n-1}$$

and

$$\sigma_B^{\mathrm{HT}}(u) = \frac{\lambda_{n-1}(B^o|u^\perp)}{\kappa_{n-1}}, \qquad u \in S^{n-1}.$$

Thus, the isoperimetrices are given by

$$\mathbb{I}_B^{\mathrm{BU}} = \frac{\kappa_{n-1}}{\kappa_n} \lambda_n(B)\, \mathrm{I}^o B, \tag{25}$$

where I is the intersection body operator and $\mathrm{I}^o B := (\mathrm{I}B)^o$ is the polar intersection body of B, and

$$\mathbb{I}_B^{\mathrm{HT}} = \frac{\kappa_n}{\kappa_{n-1}} \frac{1}{\lambda(B^o)}\, \Pi B^o, \tag{26}$$

where Π is the projection body operator. For information on intersection and projection bodies, we refer to Schneider[40] and particularly to Gardner[28].

5. Nonexistence and Existence of (Positive) Crofton Measures

Crofton measures will now first be studied in a Minkowski space $(\mathbb{R}^n, \|\cdot\|)$, with unit ball B. We assume that $k \in \{2, \ldots, n-1\}$ and that a Minkowskian area defined by a k-normalization α_k satisfying **(M1)** - **(M3)**, and **(M4)** in the case $k = n-1$, is given. By a **Crofton measure for the area** α_k^B we understand a translation invariant measure φ_{n-k} on the affine Grassmannian $A(n, n-k)$ that satisfies

$$\int_{A(n,n-k)} \mathrm{card}\,(E \cap K)\, \varphi_{n-k}(\mathrm{d}E) = \alpha_k^B(K) \tag{27}$$

for every k-dimensional convex body K. Thus, in a Minkowski space, we make the translation invariance part of the definition of a Crofton measure. Further, we require the validity of the Crofton type formula (27) only for k-dimensional convex bodies K. The existence or nonexistence of Crofton measures is already decided in this simple case, and if a Crofton measure exists, then it will later be possible to prove Crofton type formulas for much more general sets.

The question for the existence of Crofton measures for general Minkowskian (and even more general affine) areas goes back to Busemann[19], and he has obtained, in a weaker form, the following criterion. Let $G(n,j)$ denote the Grassmannian of j-dimensional linear subspaces of $\mathbb{R}^d$. For $E \in G(n,k)$ and $L \in G(n,n-k)$ we denote by $[E,L]$ the absolute k-dimensional determinant of the orthogonal projection from E to $L^\perp$. A proof of the following lemma can be found in Schneider and Wieacker[48].

Lemma 5.1. *A Crofton measure for α_k^B exists if and only if there is a finite measure ψ on $G(n,n-k)$ such that*

$$\frac{\alpha_k(B \cap E)}{\lambda_k(B \cap E)} = \int_{G(n,n-k)} [E,L]\, \psi(\mathrm{d}L) \qquad \text{for } E \in G(n,k). \tag{28}$$

In the case $k = n-1$, the left-hand side of (28) is the value of the scaling function at the unit normal vector of the $(n-1)$-dimensional linear subspace E. In this case, we can replace the integration over $G(n,1)$ by an integration over the unit sphere S^{n-1}. Moreover, since we have assumed **(M4)**, the scaling function is the support function of the convex body $\mathbf{I}_B$, a multiple of the isoperimetrix. Equation (28) is thus equivalent to

$$h(\mathbf{I}_B, u) = \int_{S^{n-1}} |\langle u, v\rangle|\, \varphi(\mathrm{d}v) \qquad \text{for } u \in \mathbb{R}^n, \tag{29}$$

with a finite even measure φ on S^{n-1}. Hence, we obtain the following result. The isoperimetrix and the body $\mathrm{I}^o B$ occurring here depend on additional data (a volume, or the Euclidean metric), but only up to a factor, which is irrelevant.

Theorem 5.1. *For a Minkowskian $(n-1)$-area a Crofton measure (a positive Crofton measure) exists if and only if the isoperimetrix is a generalized zonoid (a zonoid).*

For the Busemann $(n-1)$-area a Crofton measure (a positive Crofton measure) exists if and only if $\mathrm{I}^o B$, the polar intersection body of the unit ball, is a generalized zonoid (a zonoid).

For the Holmes–Thompson $(n-1)$-area, a positive Crofton measure always exists.

The latter assertion follows from (26) and the fact that projection bodies are zonoids, hence the zonoid equation (29) for the support function of $\mathbf{I}_B$ can be solved with a positive measure φ.

By Theorem 5.1, the existence of Crofton measures for Minkowskian $(n-1)$-areas is closely connected with the theory of generalized zonoids. We will now exploit this connection.

Theorem 5.2. *There exist Minkowski spaces, for example* ℓ_∞^n *and* ℓ_1^n, *for which the only Minkowskian* $(n-1)$*-area admitting a positive Crofton measure is the Holmes–Thompson area (up to a factor).*

This was proved in Schneider[41]. By Theorem 5.1, the proof reduces to a question on zonoids, of the following type. The unit ball of the Minkowski space in question determines two centred polytopes P and Q, where $Q \subset P$, $\mu Q \not\subset P$ for $\mu > 1$, and P is a zonotope. One has to show that a zonoid Z satisfying $Q \subset Z \subset P$ necessarily coincides with P. For example, if $n = 3$, then the pair (Q, P) in the case of ℓ_∞^3 is the pair (octahedron, rhombic dodecahedron), and in the case of ℓ_1^3 the pair (cuboctahedron, cube). That assertions about general Minkowskian surface areas are possible at all, is due, roughly speaking, to Axiom **(M1)** and the fact that for the cube and the cross-polytope many central hyperplane sections are linearly equivalent.

In the following, we use the Banach–Mazur distance to topologize the set of all n-dimensional Minkowski spaces (more precisely, the set of isometry classes of n-dimensional Minkowski spaces).

Theorem 5.3. *In every Minkowski space of sufficiently large dimension* n *which is sufficiently close to* ℓ_∞^n, *there exists no positive Crofton measure for the Busemann* $(n-1)$*-area.*

The basic ideas of the proof, given in Schneider[42], are the following. According to Theorem 5.1, we have to show that for the spaces in question the polar intersection body $\mathrm{I}^o B$ of the unit ball B is not a zonoid. For this, we need a lemma expressing that cross-polytopes are very far from zonoids. Let $(e_1, \dots, e_n)$ be an orthonormal basis of $\mathbb{R}^n$; then $Q := \mathrm{conv}\{\pm e_1, \dots, \pm e_n\}$ is a cross-polytope.

Lemma 5.2. *If*

$$0 < \gamma < \gamma_n := 2^{-(n-1)} n \binom{n-1}{\lfloor \frac{n-1}{2} \rfloor}$$

and K *is a convex body with*

$$Q \subset K \subset \gamma Q,$$

then K *is not a zonoid.*

As a side remark, we point out that $\gamma_n \sim \sqrt{2n/\pi}$ for $n \to \infty$.

The space ℓ_∞^n can be identified with $\mathbb{R}^n$ with unit ball given by the cube C with vertices $\pm e_1 \pm \cdots \pm e_n$. We consider the multiple $\alpha \mathrm{I}^o C$ of the polar intersection body of C that satisfies $h(\alpha \mathrm{I}^o C, e_1) = 1$. Since $\alpha \mathrm{I}^o C$ has the same symmetries as C, it follows that

$$Q \subset \alpha \mathrm{I}^o C.$$

Suppose that also

$$\alpha \mathrm{I}^o C \subset \operatorname{int} \lambda_n Q. \tag{30}$$

If this holds, then $Q \subset \operatorname{int} a \mathrm{I}^o C \subset \operatorname{int} \lambda Q$ for suitable numbers $a > 1$ and $\lambda < \lambda_n$. Since the intersection body operator is continuous, there is a neighbourhood U of the cube such that for all Minkowskian unit balls $B \subset U$ the polar intersection body also satisfies

$$Q \subset \operatorname{int} a \mathrm{I}^o B \subset \operatorname{int} \lambda Q.$$

By Lemma 5.2, $\mathrm{I}^o B$ is not a zonoid, hence there is no positive Crofton measure for the Busemann area in the Minkowski space with unit ball B.

To establish (30), we set $z = e_1 + \cdots + e_n$ and show that

$$h(\alpha \mathrm{I}^o C, z) < \lambda_n h(Q, z). \tag{31}$$

If this is proved, then (30) follows by symmetry. Now

$$h(\alpha \mathrm{I}^o C, z) = \frac{\sqrt{n}}{S(n)},$$

where $S(n)$ denotes the $(n-1)$-volume of the intersection of the cube $\frac{1}{2}C$ with a hyperplane through its centre and orthogonal to a main diagonal. It is known (see, for example, Chakerian and Logothetti[23]) that

$$S(n) = \frac{2}{n}\sqrt{\pi} \int_0^\infty \left(\frac{\sin x}{x}\right)^n \mathrm{d}x \to \sqrt{6/\pi}$$

for $n \to \infty$. From this, it follows that (31) is satisfied for all sufficiently large dimensions n.

Theorem 5.4. *There exist Minkowski spaces arbitrarily close to the Euclidean space ℓ_2^n in which there exists no positive Crofton measure for the Busemann $(n-1)$-area.*

To obtain this result, one has to construct convex bodies, arbitrarily close to the Euclidean unit ball B^n, for which the polar intersection body is not a zonoid. In Schneider[42], this is achieved as follows. Let $u, z \in S^{n-1}$ be orthogonal unit vectors, and let $\epsilon > 0$. Define

$$B_0 := \operatorname{conv}(B^n \cup (1+\epsilon)(B^n \cap u^\perp))$$

and $B := B_0 + \epsilon \operatorname{conv}\{-z, z\}$. By computing the directional derivatives of the section volume function $v \mapsto \lambda_{n-1}(B \cap v^\perp)$ at u, one can deduce that the face $F(\mathrm{I}^o B, u)$ of the polar intersection body of B with outer normal vector u contains an $(n-1)$-dimensional ball as a summand.

Further, the body B has a cylindrical part. This implies that there is a neighbourhood U of the vector z such that

$$h(\mathrm{I}^o B, y) = h(\mathrm{I}^o B, z)\langle y, z\rangle \qquad \text{for } y \in U.$$

This means that the body $\mathrm{I}^o B$ has a vertex z_0 with outer normal vector z.

Now assume that $\mathrm{I}^o B$ were a zonoid. Then the face $F(\mathrm{I}^o B, u)$ is a summand of $\mathrm{I}^o B$. In particular, $\mathrm{I}^o B$ has a summand K which is an $(n-1)$-dimensional ball. There is a translate K' of K such that $z_0 \in K' \subset \mathrm{I}^o B$. But this is not possible, since z_0 is a vertex of $\mathrm{I}^o B$. Thus $\mathrm{I}^o B$ cannot be a zonoid.

In general, it is difficult to verify that a given convex body is not a zonoid, except in the trivial case where it has a face that is not centrally symmetric. This difficulty is one obstacle for a proof of the following:

Conjecture. In the space of n-dimensional Minkowski spaces, there is a dense subset of spaces in which there is no positive Crofton measure for the Busemann $(n-1)$-area.

On the other hand, it would be rash to conjecture that a positive Crofton measure for the Busemann area existed only in Euclidean spaces.

Theorem 5.5. *There exist Minkowski spaces arbitrarily close to ℓ_2^n, but not Euclidean, in which there does exist a positive Crofton measure for the Busemann $(n-1)$-area.*

This is proved in Schneider[42], by smooth perturbation of the Euclidean unit ball B^n. It is shown that this can be done in such a way that the obtained body B_ϵ is convex, centrally symmetric and smooth, but not an ellipsoid, and that for the support function of the polar intersection body $\mathrm{I}^o B_\epsilon$, the zonoid equation still has a positive solution. This means that $\mathrm{I}^o B_\epsilon$

is a zonoid, hence in the Minkowski space with unit ball B_ϵ, the Busemann $(n-1)$-area admits a positive Crofton measure.

Theorem 5.5 gives a positive answer to the third of the open problems in Chakerian[22].

One consequence of the preceding results is the conclusion that the Busemann $(n-1)$-area, although very natural, being a Hausdorff measure, is not suitable for integral geometry, since for it not even the simplest Crofton formulas with positive measures exist in all Minkowski spaces. Even more restrictions arise in Finsler spaces. In a Minkowski space, under strong smoothness assumptions, the zonoid equation (29) for the support function of the isoperimetrix $\mathbb{I}_B^{\mathrm{BU}}$ has a solution, hence there exists a signed Crofton measure for the Busemann $(n-1)$-area. In a projective Finsler space, even smoothness assumptions are not sufficient to obtain Crofton formulas for the Busemann area. An example to this effect was constructed by Álvarez and Berck[6].

For the Holmes–Thompson area, the situation is much better. This is already seen from the last part of Theorem 5.1, asserting that in every n-dimensional Minkowski space a positive Crofton measure exists for the Holmes–Thompson $(n-1)$-area vol_{n-1}. For the lower-dimensional Holmes–Thompson areas vol_k, the following holds.

Theorem 5.6. *If in an n-dimensional Minkowski space there exists a Crofton measure (a positive Crofton measure) for the norm, then there also exists a Crofton measure (a positive Crofton measure) for* vol_k, $k = 2, \dots, n-2$.

This follows from the first part of Theorem 2.1 and the construction in Section 7. There are two main cases where the assumption of Theorem 5.6 is satisfied:

- If the norm $\|\cdot\| = h(B^o, \cdot)$ is sufficiently smooth, then B^o is a generalized zonoid, hence a Crofton measure for vol_1 exists.
- If the Minkowski space $(\mathbb{R}, \|\cdot\|)$ is hypermetric then, by Theorem 2.2, a positive Crofton measure for vol_1 exists.

Under either of these two assumptions, smooth or hypermetric, the existence of Crofton measures for the Holmes–Thompson areas of all dimensions extends to projective Finsler spaces, and Crofton formulas for quite general subsets can be proved. For this, more information on generalized zonoids is helpful, and this will be collected in the next section.

6. More on Generalized Zonoids

Let $Z \subset \mathbb{R}^n$ be a generalized zonoid with centre 0. Thus, the support function of Z has an integral representation

$$h(Z,\xi) = \int_{S^{n-1}} |\langle \xi, u\rangle| \, \rho(\mathrm{d}u) \tag{32}$$

with a finite (signed) measure ρ on the sphere S^{n-1}. This equation can be interpreted as giving half the one-dimensional volume of the orthogonal projection of Z on to the linear subspace spanned by ξ. There is an extension to volumes of higher-dimensional projections. By $L(u_1,\dots,u_k)$ and $[u_1,\dots,u_k]$ we denote, respectively, the linear subspace spanned by the vectors $u_1,\dots,u_k$ and the k-dimensional (Euclidean) volume of the parallelepiped spanned by these vectors. Let $k \in \{1,\dots,n\}$ and $E \in A(n,k)$. Then

$$\begin{aligned} &\lambda_k(Z|E) \\ &= \frac{2^k}{k!} \int_{S^{n-1}} \cdots \int_{S^{n-1}} [E, L(u_1,\dots,u_k)^\perp] \, [u_1,\dots,u_k] \, \rho(\mathrm{d}u_1)\cdots\rho(\mathrm{d}u_k). \end{aligned} \tag{33}$$

The proof given by Weil[52] holds also for signed measures. Equation (33) can be written in a more concise form, after defining the 'projection generating measure' $\rho^{(k)}$ on $G(n,k)$ by

$$\begin{aligned} &\rho^{(k)}(A) \\ &:= c_k \int_{S^{n-1}} \cdots \int_{S^{n-1}} \mathbf{1}_A(L(u_1,\dots,u_k))[u_1,\dots,u_k] \, \rho(\mathrm{d}u_1)\cdots\rho(\mathrm{d}u_k) \end{aligned} \tag{34}$$

for Borel sets $A \subset G(n,k)$, with c_k given by

$$c_k := \frac{2^k}{k!\kappa_k}.$$

Then (33) takes the form

$$\lambda_k(Z|E) = \kappa_k \int_{G(n,k)} [E, L^\perp] \, \rho^{(k)}(\mathrm{d}L) \qquad \text{for } E \in G(n,k). \tag{35}$$

The definition (34) essentially goes back to Matheron[35], p. 101; later uses of the projection generating measure begin with Goodey and Weil[30].

If we define $\rho_{(n-k)}$ as the image measure of $\rho^{(k)}$ under the map $L \mapsto L^\perp$ from $G(n,k)$ to $G(n,n-k)$, then (35) can be written in the form

$$\frac{\lambda_k(Z|E)}{\kappa_k} = \int_{G(n,n-k)} [E, L] \, \rho_{(n-k)}(\mathrm{d}L) \qquad \text{for } E \in G(n,k). \tag{36}$$

The case $k = n-1$ has a special feature, since the measure $\rho_{(n-1)}$ is related to the area measure $S_{n-1}(Z,\cdot)$ of Z. For $u \in S^{n-1}$ we have, by (36) and a well-known representation of the projection volume,

$$\frac{\lambda_k(Z|u^\perp)}{\kappa_{n-1}} = \int_{G(n,1)} [u^\perp, L]\,\rho_{(1)}(\mathrm{d}L) = \frac{1}{2\kappa_{n-1}} \int_{S^{n-1}} |\langle u, v\rangle|\, S_{n-1}(Z, \mathrm{d}v).$$

From the uniqueness result for the zonoid equation, it follows that the measure $2\kappa_{n-1}\rho_{(1)}$ is the image measure of $S_{n-1}(Z,\cdot)$ under the map $u \mapsto L(u)$ from S^{n-1} to $G(n,1)$.

Now we assume that Z has a representation

$$h(Z,\xi) = \int_{S^{n-1}} |\langle \xi, u\rangle| g(u)\,\sigma(\mathrm{d}u)$$

with a continuous function g. Let $s_{n-1}(Z,u)$ be the product of the principal radii of curvature of Z at the boundary point with outer unit normal vector u. The following formula, proved by Weil[51], Satz 7, will be needed:

$$s_{n-1}(Z,u) = \frac{2^{n-1}}{(n-1)!} \int_{s_u} \cdots \int_{s_u} [u_1, \ldots, u_{n-1}]^2 g(u_1) \cdots g(u_{n-1}) \times \sigma_u(\mathrm{d}u_1) \cdots \sigma_u(\mathrm{d}u_{n-1}). \tag{37}$$

Here σ_u is the $(n-2)$-dimensional spherical Lebesgue measure on the sphere $s_u := S^{n-1} \cap u^\perp$.

7. Crofton Formulas in Smooth Projective Finsler Spaces

The results on generalized zonoids have immediate applications to the existence of Crofton measures for Holmes-Thompson areas. Let us first consider a Minkowski space $(\mathbb{R}^n, \|\cdot\|)$ for which the polar unit ball B^o is a generalized zonoid. Then there is a representation

$$h(B^o, \xi) = \int_{S^{n-1}} |\langle \xi, u\rangle|\, \rho(\mathrm{d}u) \tag{38}$$

with a finite measure ρ. For $k \in \{1, \ldots, n-1\}$, let $\rho_{(n-k)}$ be the corresponding measure on $G(n, n-k)$ as defined in Section 6. By Lemma 5.1, (36) is precisely the condition which ensures that a Crofton measure exists for the Holmes–Thompson k-area vol_k. This Crofton measure η_{n-k} is defined by

$$\int_{A(n,n-k)} f\,\mathrm{d}\eta_{n-k} := \int_{G(n,n-k)} \int_{L^\perp} f(L+t)\,\lambda_k(\mathrm{d}t)\,\rho_{(n-k)}(\mathrm{d}L) \tag{39}$$

for nonnegative measurable functions f on $A(n, n-k)$. In terms of parameterized hyperplanes, it can be represented by

$$\int_{A(n,n-k)} f \, \mathrm{d}\eta_{n-k} = c_k \int_{S^{n-1}} \cdots \int_{S^{n-1}} \int_{\mathbb{R}} \cdots \int_{\mathbb{R}} f(H_{u_1,t_1} \cap \cdots \cap H_{u_k,t_k}) \times \mathrm{d}t_1 \cdots \mathrm{d}t_k \, \rho(\mathrm{d}u_1) \cdots \rho(\mathrm{d}u_k).$$

Thus, η_{n-k} is the image measure of $c_k(\eta_{n-1})^{\otimes k}$ under the intersection map

$$(H_1, \ldots, H_k) \mapsto H_1 \cap \cdots \cap H_k$$

from the set of independent k-tuples of hyperplanes to $A(n, n-k)$.

This existence and representation of Crofton measures for Holmes–Thompson areas was first obtained in Schneider and Wieacker[48], for hypermetric Minkowski spaces. In that case, the Crofton measures are positive. The proofs given in Schneider and Wieacker[48] carry over, without change, to Minkowski spaces whose polar unit balls are generalized zonoids, thus, in particular, to spaces with sufficiently smooth norms. The Crofton measures are then signed measures. However, the line measure η_1 is always a positive measure. This follows from the remark in Section 6 concerning $\rho_{(1)}$. It shows that the line measure η_1 given by (39) can also be represented by

$$\int_{A(n,1)} f \, \mathrm{d}\eta_1 = \frac{1}{2\kappa_{n-1}} \int_{S^{n-1}} \int_{u^\perp} f(L(u) + x) \, \lambda_{n-1}(\mathrm{d}x) \, S_{n-1}(B^o, \mathrm{d}u) \quad (40)$$

and thus is positive. This measure on the space of lines can be defined in general Minkowski spaces; it appeared already in Busemann[19] and later in El-Ekhtiar[25].

The general Crofton formulas obtained in Schneider and Wieacker[48], and the methods to prove them, extend to smooth projective Finsler spaces, and will now be formulated in this generality.

Let $(\mathbb{R}^n, F)$ be a smooth projective Finsler space. We combine Pogorelov's approach with the preceding one. In particular, the representation (19) of the local polar unit ball B_x^o at $x \in \mathbb{R}^n$ replaces now the representation (38). The (signed) measure $\eta_{n-1} = \eta$ is defined by (18). Following the procedure above, for $k \in \{2, \ldots, n-1\}$ we define η_{n-k} as the image measure of $c_k(\eta_{n-1})^{\otimes k}$ under the intersection map. Explicitly, this means that

$$\int_{A(n,n-k)} f \, \mathrm{d}\eta_{n-k} = c_k \int_{S^{n-1}} \cdots \int_{S^{n-1}} \int_{\mathbb{R}} \cdots \int_{\mathbb{R}} f(H_{u_1,t_1} \cap \cdots \cap H_{u_k,t_k}) \times g(u_1, t_1) \cdots g(u_k, t_k) \, \mathrm{d}t_1 \cdots \mathrm{d}t_k \, \sigma(\mathrm{d}u_1) \cdots \sigma(\mathrm{d}u_k). \quad (41)$$

With these Crofton measures, we can formulate general Crofton formulas for Holmes–Thompson areas. They are general in two respects: the submanifolds need not be smooth, and submanifold and intersecting flats need not be of complementary dimensions.

In the following, we denote by $\mathcal{H}^k$ the k-dimensional Hausdorff measure that is induced by the auxiliary Euclidean metric on $\mathbb{R}^n$, for $k \in \{0, \ldots, n\}$. The restriction of this outer measure to the Borel sets is σ-additive. A set $M \subset \mathbb{R}^n$ is called **$(\mathcal{H}^k, k)$-rectifiable** if $\mathcal{H}^k(M) < \infty$ and there exist Lipschitz maps $f_i : \mathbb{R}^k \to \mathbb{R}^n$, $i \in \mathbb{N}$, such that $\mathcal{H}^k(K \setminus \bigcup_{i\in\mathbb{N}} f_i(\mathbb{R}^k)) = 0$. This notion does not depend on the choice of the Euclidean metric. For a $(\mathcal{H}^k, k)$-rectifiable Borel set, we can define the k-dimensional Holmes–Thompson area by extending (23),

$$\mathrm{vol}_k(M) := \frac{1}{\kappa_k} \int_M \lambda_k(B_x^o | T_x M)\, \mathcal{H}^k(\mathrm{d}x). \tag{42}$$

The approximate tangent space $T_x M$ exists and is unique for $\mathcal{H}^k$-almost all $x \in M$; it is a measurable function of x.

With these definitions, a general Crofton formula of type (1) can be obtained. Moreover, it can be extended to the case where the submanifold and the intersecting flats are no longer of complementary dimensions.

Theorem 7.1. *Let $\mathbb{R}^n$ be endowed either with a norm for which the polar unit ball is a generalized zonoid, or with a smooth projective Finsler metric. Let $M \subset \mathbb{R}^n$ be a $(\mathcal{H}^k, k)$-rectifiable Borel set, $k \in \{1, \ldots, n\}$. Then*

$$\int_{A(n,n-k)} \mathrm{card}\,(E \cap M)\, \eta_{n-k}(\mathrm{d}E) = \mathrm{vol}_k(M). \tag{43}$$

More generally, if $j \in \{n-k, \ldots, n-1\}$, then

$$\int_{A(n,j)} \mathrm{vol}_{k+j-n}(E \cap M)\, \eta_j(\mathrm{d}E) = \frac{c_{k+j-n} c_{n-j}}{c_k} \mathrm{vol}_k(M). \tag{44}$$

Theorem 7.1 can be viewed as giving a positive answer, for projective Finsler spaces, to the first of the three open problems formulated by Chakerian[22].

As mentioned, formula (44) for hypermetric Minkowski spaces was first proved in Schneider and Wieacker[48]. Formula (43) and the case $k = n-1$ of (44) for smooth submanifolds of smooth projective Finsler spaces are due to Álvarez and Fernandes[7]. Their proof employs the symplectic structure on the space of geodesics of a projective Finsler space. The theme of Crofton formulas in smooth projective Finsler spaces was taken up in Álvarez and

Fernandes[8] and in the thesis of Fernandes[26], where now part of the methods is closer to those in Schneider and Wieacker[48], in so far as connections to Fourier transforms of norms, and thus to cosine transforms and the zonoid equation, are used. In Fernandes[26] and in Álvarez and Fernandes[10], the topic is considerably expanded, and the role of double fibrations and the Gelfand transform is emphasized. This approach to Crofton formulas is neatly set out in the survey article by Álvarez and Fernandes[9].

The general formula (44) for smooth projective Finsler spaces was proved in Schneider[44].

About the line measure η_1 we remark that the representation (40) has now the counterpart

$$\int_{A(n,1)} f \,\mathrm{d}\eta_1 = \frac{1}{2\kappa_{n-1}} \int_{S^{n-1}} \int_{u^\perp} f(L(u)+x) s_{n-1}(B_x^o, u)\, \lambda_{n-1}(\mathrm{d}x)\, \sigma(\mathrm{d}u). \tag{45}$$

The proof uses (37) and is found in Schneider[43], p. 95. In other words, a density δ of the line measure η_1 with respect to the motion invariant measure on $A(n,1)$ is given by $\delta(L(u)+x) = s_{n-1}(B_x^o, u)$. Since this density is nonnegative, η_1 is again a positive measure.

The form of the density δ has the following consequence. If x and y lie on a line with direction u, then $L(u)+x = L(u)+y$ and hence $s_{n-1}(B_x^o, u) = s_{n-1}(B_y^o, u)$. This should not come as a surprise; it is implied by the remark following Theorem 3.1.

8. Crofton Measures in General Projective Finsler Spaces

Crofton measures for Holmes–Thompson areas in general Finsler spaces, without smoothness assumptions, have so far only been found in cases where the measures turn out to be positive.

Let $(\mathbb{R}^n, F)$ be a hypermetric projective Finsler space. By Theorem 3.2, due to Alexander, there exists a positive measure η on the space $A(n, n-1)$ of hyperplanes such that

$$\mathrm{vol}_1(S) = \int_{A(n,n-1)} \mathrm{card}(E \cap S)\, \eta(\mathrm{d}E) \tag{46}$$

for every line segment S; here vol_1 is the curve length in the Finsler space. This result can be extended to the higher dimensional Holmes–Thompson areas, with positive Crofton measures. Let $k \in \{1, \dots, n-1\}$. Motivated by the procedure of Schneider and Wieacker[48] in hypermetric Minkowski spaces, we define a positive measure η_{n-k} on the space $A(n, n-k)$ of $(n-k)$-

flats as the image measure, under the intersection map

$$(H_1, \dots, H_k) \mapsto H_1 \cap \dots \cap H_k,$$

of the product measure $c_k \eta^{\otimes k}$, restricted to the set of k-tuples $(H_1, \dots, H_k)$ with $(n-k)$-dimensional intersection.

Theorem 8.1. *For $k \in \{1, \dots, n-1\}$, the positive measure η_{n-k}, defined in the hypermetric projective Finsler space $(\mathbb{R}^n, F)$, satisfies*

$$\mathrm{vol}_k(M) = \int_{A(n,n-1)} \mathrm{card}(E \cap M)\, \eta_{n-k}(\mathrm{d}E) \tag{47}$$

for every k-dimensional compact convex set M.

This was proved in Schneider[43]. Theorem 8.1 gives an answer to the second open problem of Chakerian[22]. The extension to more general sets M and to formulas of type (44) has not been investigated. The validity of (47) for k-dimensional compact convex sets already suffices to obtain the uniqueness of the measure η_{n-k} for $k = 1$ (shown in Alexander[2]) and $k = n-1$ (shown in Schneider[43]). For $k \in \{2, \dots, n-2\}$, uniqueness fails.

The proof of Theorem 8.1 uses approximation. By results of Pogorelov[36] and Szabó[49], there is for every $\epsilon > 0$ a Finsler metric F_ϵ on $\mathbb{R}^n$ such that $(\mathbb{R}^n, F_\epsilon)$ is a smooth projective Finsler space and $\lim_{\epsilon \to 0} F_\epsilon = F$ uniformly on every compact set. The Finsler metric F_ϵ is hypermetric, too (as noted by Alexander), hence every norm $F_\epsilon(x, \cdot)$ is hypermetric. Therefore, every local polar unit ball B_x^o is a zonoid, which implies that the function g appearing in (18) is nonnegative. Hence, the measure η_ϵ defined by (18) for the smooth Finsler metric F_ϵ is positive. The same holds for the derived measures $(\eta_\epsilon)_{n-k}$. One can then let ϵ tend to 0 and work with vaguely converging subsequences of the corresponding sequences of positive measures to complete the proof of Theorem 8.1.

The good properties of the hypermetric projective Finsler spaces raise the following open question:

Problem. Are the only hypermetric Hilbert geometries of dimension $n > 2$ the hyperbolic geometries? Equivalently: Let K be an n-dimensional convex body ($n > 2$) with the property that $(K-x)^o - (K-x)^o$ is a zonoid for every $x \in \mathrm{int}\, K$. Must K be an ellipsoid?

The measure $(\eta_\epsilon)_1$ constructed above is always positive, by (45), even if the spaces are not hypermetric. Therefore, the approximation procedure can be modified to yield the following result (details are in Schneider[43]).

Theorem 8.2. *In a general projective Finsler space* $(\mathbb{R}^n, F)$, *there always exists a unique positive measure* η_1 *on the space* $A(n,1)$ *of lines such that*

$$\mathrm{vol}_k(M) = \int_{A(n,n-1)} \mathrm{card}(E \cap M)\, \eta_1(\mathrm{d}E)$$

for every $(n-1)$*-dimensional compact convex set* M.

The existence and uniqueness of the positive line measure η_1 can be extended, without essential changes of the method, to general projective Finsler spaces (C, F), where $C \subset \mathbb{R}^n$ is a bounded open convex subset. For Minkowski spaces, the line measure is explicitly given by (40). This case includes also non-smooth projective Finsler spaces. For example, one can read off from (40) on which sets of lines the line measure is concentrated if the unit ball is a polytope. Besides Minkowski spaces, the only non-smooth cases where the line measure η_1 is explicitly known are the Hilbert geometries in a convex polygon in the plane; see Alexander[2] and, in more detail, Alexander, Berg and Foote[3]. The higher-dimensional case requires a different approach. In Schneider[45], the case of the Hilbert geometry in an n-dimensional polytope is investigated. It turns out that the line measure is concentrated on the set of lines meeting two disjoint faces of the polytope whose dimensions add up to $n-1$. This set of lines has dimension $n-1$, whereas $A(n,1)$ itself has dimension $2(n-1)$.

References

1. R. Alexander, Planes for which the lines are the shortest paths between points. *Illinois J. Math.* **22** (1978), 177–190.
2. R. Alexander, Zonoid theory and Hilbert's fourth problem. *Geom. Dedicata* **28** (1988), 199–211.
3. R. Alexander, I.D. Berg, R. Foote, Integral-geometric formulas for perimeter in $\mathbb{S}^2$, $\mathbb{H}^2$, and Hilbert planes. *Rocky Mountain J. Math.* (to appear).
4. J.C. Álvarez Paiva, Hilbert's fourth problem in two dimensions I. In: "Mass Selecta: Teaching and Learning Advanced Undergraduate Mathematics" (S. Katok, A. Sossinsky, S. Tabachnikov, eds.), pp. 165–183, Amer. Math. Soc., Providence, RI 2003.
5. J.C. Álvarez Paiva, Symplectic geometry and Hilbert's fourth problem. *J. Differential Geom.* (to appear).
6. J.C. Álvarez Paiva, G. Berck, What is wrong with the Hausdorff measure in Finsler spaces. Preprint, arXiv:math.DG/0408413
7. J.C. Álvarez Paiva, E. Fernandes, Crofton formulas in projective Finsler spaces. *Electron. Res. Announc. Amer. Math. Soc.* **4** (1998), 91–100.

8. J.C Álvarez Paiva, E. Fernandes, Fourier transforms and the Holmes–Thompson volume of Finsler manifolds. *Int. Math. Res. Not.* **19** (1999), 1031–1042.
9. J.C Álvarez Paiva, E. Fernandes, What is a Crofton formula? *Math. Notae* **42** (2003/04), 95–108.
10. J.C. Álvarez Paiva, E. Fernandes, Gelfand transforms and Crofton formulas. *Selecta Math.* (to appear).
11. J.C. Álvarez Paiva, I.M. Gelfand, M.M. Smirnov, Crofton densities, symplectic geometry, and Hilbert's fourth problem. In: "Arnold–Gelfand Mathematical Seminars, Geometry and Singularity Theory" (V.I. Arnold, I.M. Gelfand, V.S. Retakh, eds.), pp. 77–92, Birkhäuser, Boston 1997.
12. J.C. Álvarez Paiva, A.C. Thompson, Volumes on normed and Finsler spaces. In: "A Sampler of Riemann–Finsler Geometry" (D. Bao, R.L. Bryant, S.-S. Chern, Z. Shen, eds.), pp. 1–48, MSRI Publ., vol. 50, Cambridge University Press 2004.
13. R.V. Ambartzumian, A note on pseudo-metrics in the plane. *Z. Wahrscheinlichkeitsth. verw. Geb.* **37** (1976), 145–155.
14. R.V. Ambartzumian, V.K. Oganian, Parametric versions of Hilbert's fourth problem. *Israel J. Math.* **103** (1998), 41–65.
15. G. Bellettini, M. Paolini, S. Venturini, Some results on surface measures in calculus of variations. *Ann. Mat. Pura Appl.* (4) **170** (1996), 329 - 357.
16. W. Blaschke, Über die Figuratrix in der Variationsrechnung. *Arch. Math. Phys.*, III. Reihe, **20** (1913), 45–61.
17. E.D. Bolker, A class of convex bodies. *Trans. Amer. Math. Soc.* **145** (1969), 323–346.
18. H. Busemann, The isoperimetric problem for Minkowski area. *Amer. J. Math.* **71** (1949), 743–762.
19. H. Busemann, Areas in affine spaces III. The integral geometry of affine area. *Rend. Circ. Mat. Palermo* (2) **9** (1960), 226–242.
20. H. Busemann, Geometries in which the planes minimize area. *Ann. Mat. Pura Appl.* (4) **55** (1961), 171–190.
21. H. Busemann, Problem IV: Desarguesian Spaces. *Proc. A.M.S. Symp. Pure Math.* **28** (1976), 131–141.
22. G.D. Chakerian, Integral geometry in Minkowski spaces. In: "Finsler Geometry" (D. Bao et al., eds.), pp. 43–50; *Contempory Math.* **196**, Amer. Math. Soc., Providence, RI 1996.
23. G.D. Chakerian, D. Logothetti, Cube slices, pictorial triangles, and probability. *Math. Mag.* **64** (1991), 219–241.
24. M.M. Deza, M. Laurent, *Geometry of Cuts and Metrics.* Springer, Berlin 1997.
25. A. El-Ekhtiar, Integral geometry in Minkowski space. Ph.D. Thesis, University of California, Davis 1992.
26. E. Fernandes, Double fibrations: a modern approach to integral geometry and Crofton formulas in projective Finsler spaces. Ph.D. Thesis, Louvain-la-Neuve 2002.
27. P. Funk, Über eine geometrische Anwendung der Abelschen Integralgleich-

ung. *Math. Ann.* **77** (1916), 129–135.

28. R.J. Gardner, *Geometric Tomography.* Encyclopedia of Math. and Its Applications, vol. 58, Cambridge University Press, Cambridge 1995.
29. I.M. Gelfand, M.M. Smirnov, Lagrangians satisfying Crofton formulas, Radon transforms, and nonlocal differentials. *Adv. Math.* **109** (1994), 188–227.
30. P. Goodey, W. Weil, Translative integral formulae for convex bodies. *Aequationes Math.* **34** (1987), 64–77.
31. P. Goodey, W. Weil, Zonoids and generalizations. In: "Handbook of Convex Geometry", vol. B (P.M. Gruber, J.M. Wills, eds.), pp. 1297–1326. North-Holland, Amsterdam 1993.
32. H. Groemer, *Geometric Applications of Fourier Series and Spherical Harmonics.* Encyclopedia of Math. and Its Applications, vol. 61, Cambridge University Press, Cambridge 1996.
33. G. Hamel, Über die Geometrien, in denen die Geraden die Kürzesten sind. *Math. Ann.* **57** (1903), 231–264.
34. J.B. Kelly, Hypermetric spaces. In: "The Geometry of Metric and Linear Spaces" (L.M. Kelly, ed.), pp. 19–31. *Lecture Notes in Math.* **490**, Springer, Berlin 1977.
35. G. Matheron, *Random Sets and Integral Geometry.* Wiley, New York 1975.
36. A.V. Pogorelov, *Hilbert's Fourth Problem.* Scripta Series in Mathematics. V.H. Winston & Sons, Washington, D.C.; A Halstead Press Book, John Wiley & Sons, New York 1979 (Russian Original: Izdat. "Nauka", Moscow 1974).
37. L.A. Santaló, *Integral Geometry and Geometric Probability.* Encyclopedia of Math. and Its Applications, vol. 1, Addison-Wesley, Reading, MA 1976.
38. R. Schneider, Zu einem Problem von Shephard über die Projektionen konvexer Körper. *Math. Z.* **191** (1967), 71–82.
39. R. Schneider, Über eine Integralgleichung in der Theorie der konvexen Körper. *Math. Nachr.* **44** (1970), 55–75.
40. R. Schneider, *Convex Bodies: the Brunn-Minkowski Theory.* Encyclopedia of Math. and Its Applications, vol. 44, Cambridge University Press, Cambridge 1993.
41. R. Schneider, On areas and integral geometry in Minkowski spaces. *Beiträge Algebra Geom.* **38** (1997), 73–86.
42. R. Schneider, On the Busemann area in Minkowski spaces. *Beiträge Algebra Geom.* **42** (2001), 263–273.
43. R. Schneider, Crofton formulas in hypermetric projective Finsler spaces. *Arch. Math.* **77** (2001), 85–97.
44. R. Schneider, On integral geometry in projective Finsler spaces. *Izv. Nats. Akad. Nauk Armenii Mat.* **37** (2002), 34–51.
45. R. Schneider, Crofton measures in polytopal Hilbert geometries. (in preparation).
46. R. Schneider, W. Weil, Zonoids and related topics. In: "Convexity and Its Applications" (P.M. Gruber, J.M. Wills, eds.), pp. 296–317, Birkhäuser, Basel 1983.

47. R. Schneider, W. Weil, *Stochastische Geometrie.* Teubner, Stuttgart 2000.
48. R. Schneider, J.A. Wieacker, Integral geometry in Minkowski spaces. *Adv. Math.* **129** (1997), 222–260.
49. Z.I. Szabó, Hilbert's Fourth Problem, I. *Adv. Math.* **59** (1986), 185–301.
50. A.C. Thompson, *Minkowski Geometry.* Encyclopedia of Math. and Its Applications, vol. 63, Cambridge University Press, Cambridge 1996.
51. W. Weil, Kontinuierliche Linearkombination von Strecken. *Math. Z.* **148** (1976), 71–84.
52. W. Weil, Centrally symmetric convex bodies and distributions, II. *Israel J. Math.* **32** (1979), 173–182.

RANDOM METHODS IN APPROXIMATION OF CONVEX BODIES

CARSTEN SCHÜTT
Department of Mathematics
University of Kiel
D-24098 Kiel, Germany
E-mail: schuett@math.uni-kiel.de

We discuss some results and methods concerning the approximation of convex bodies by polytopes with a restricted number of vertices or facets.

A convex body K in $\mathbb{R}^n$ is a compact, convex subset of $\mathbb{R}^n$ with non-empty interior and a polytope P in $\mathbb{R}^n$ is the convex hull of finitely many points in $\mathbb{R}^n$. As norm on $\mathbb{R}^n$ we consider the Euclidean norm

$$\|x\| = \left(\sum_{i=1}^{n} |x_i|^2 \right)^{\frac{1}{2}} .$$

The n-dimensional Euclidean ball with center x and radius r is denoted by $B_2^n(x,r)$, the unit ball with center 0 is denoted by B_2^n.

The convex hull of points $x_1, \ldots, x_N$ is denoted by $[x_1, \ldots, x_N]$. $\mu_{\partial K}$ is the surface area measure on ∂K. It equals the restriction of the $n-1$-dimensional Hausdorff measure to ∂K.

$\kappa_{\partial K}(x)$ is the (generalized) Gauß-Kronecker curvature at x. We write in short $\kappa(x)$ if it is clear which is the body K involved. If K has a C^2-boundary then the generalized curvature coincides with the curvature. By a result of Aleksandrov [1] the generalized Gauß-Kronecker curvature exists almost everywhere. See also [2,3,6].

There are a number of metrics that are used to measure the distance between two convex bodies C and K, the two most important are the symmetric difference metric

$$d_s(C,K) = \mathrm{vol}_n(C \triangle K) = \mathrm{vol}_n((C \setminus K) \cup (K \setminus C))$$

and the Hausdorff metric

$$d_H(C,K) = \max\left\{\max_{x\in\partial C}\min_{y\in\partial K}\|x-y\|, \max_{y\in\partial K}\min_{x\in\partial C}\|x-y\|\right\}.$$

In the case of inscribed polytopes

$$d_S(K,P) = \mathrm{vol}_n(K) - \mathrm{vol}_n(P)$$

and in the case of circumscribed polytopes

$$d_S(K,P) = \mathrm{vol}_n(P) - \mathrm{vol}_n(K).$$

We ask how

$$\inf\{d_S(K,P) | P \subseteq K \text{ and } P \text{ has } N \text{ vertices}\}$$

and

$$\inf\{d_S(K,P) | P \supseteq K \text{ and } P \text{ has } N\ (n-1)\text{-dimensional faces}\}$$

depend on the dimension n, the number of vertices N and the body K.

Theorem 1. *There is a constant $c_1 > 0$ such that for every convex body K in $\mathbb{R}^n$ and for every $N \in \mathbb{N}$ there exists a polytope $P \subseteq K$ with N vertices such that*

$$vol_n(K) - vol_n(P) \leq c_1 \frac{n\ vol_n(K)}{N^{\frac{2}{n-1}}}.$$

The constant c_1 can be chosen to be

$$\frac{8}{3}\left(\frac{vol_{n-1}(\partial B_2^n)}{vol_{n-1}(B_2^{n-1})}\right)^{\frac{2}{n-1}}.$$

This estimate is optimal, namely, there is a constant c_2 such that for all $n \in \mathbb{N}$ and all polytopes that are contained in B_2^n and have at most N vertices

$$c_2 \frac{n\ vol_n(B_2^n)}{N^{\frac{2}{n-1}}} \leq vol_n(B_2^n) - vol_n(P)$$

The estimate is due to Bronshteyn and Ivanov [5]. The optimality was shown by Gordon, Reisner and Schütt [7,8]. This also shows that roughly speaking each convex body can be approximated better by a polytope than the Euclidean ball. A refinement and simplification can be found in [13,14]. There, in order to estimate $d_S(B_2^n, P)$ from above, the following result is used. For a convex body K and a probability measure $\mathbb{P}$ on the boundary ∂K

we define the expected volume of a random polytope of N points chosen randomly from the boundary ∂K as

$$\mathbb{E}(\partial K, \mathbb{P}, N) = \int_{\partial K} \cdots \int_{\partial K} \mathrm{vol}_n\left([x_1, \ldots, x_N]\right) d\mathbb{P}(x_1) \cdots d\mathbb{P}(x_N)$$

Proposition 2. [16] *Let $\mathbb{P}$ be the normalized surface measure on ∂B_2^n. Then we have*

$$\lim_{N \to \infty} \frac{vol_n(B_2^n) - \mathbb{E}(\partial B_2^n, \mathbb{P}, N)}{N^{-\frac{2}{n-1}}}$$
$$= \frac{vol_{n-2}(\partial B_2^{n-1})}{2(n+1)!} \left(\frac{(n-1) vol_{n-1}(\partial B_2^n)}{vol_{n-2}(\partial B_2^{n-1})} \right)^{\frac{n+1}{n-1}} \Gamma\left(n + 1 + \frac{2}{n-1}\right)$$

This result gives a good estimate from above for

$$\inf\{\mathrm{vol}_n(B_2^n \setminus P) | P \subseteq B_2^n \text{ and } P \text{ has at most N vertices}\}$$

By a result of Macbeath [12] for all convex bodies K with $\mathrm{vol}_n(K) = \mathrm{vol}_n(B_2^n)$

$$\inf\{\mathrm{vol}_n(K \setminus P) | P \subseteq K \text{ and } P \text{ has at most N vertices}\}$$
$$\leq \inf\{\mathrm{vol}_n(B_2^n \setminus P) | P \subseteq B_2^n \text{ and } P \text{ has at most N vertices}\}$$

and thus we get a good estimate from above for all convex bodies. The corresponding estimate from below can be found in [13,14].

It is interesting to ask how much better the estimates get if we drop the condition that P is contained in B_2^n.

Theorem 3. [11] *There is a constant c such that for every $n \in \mathbb{N}$ there is a N_n so that for every $N \geq N_n$ there is a polytope P in $\mathbb{R}^n$ with N vertices such that*

$$vol_n(B_2^n \triangle P) \leq c \ vol_n(B_2^n) N^{-\frac{2}{n-1}}. \tag{1}$$

It has been shown in [4] that for a polytope P with at most N vertices

$$\mathrm{vol}_n(B_2^n \triangle P) \geq \frac{1}{67 e^2 \pi} \frac{1}{n} \mathrm{vol}_n(B_2^n) N^{-\frac{2}{n-1}}. \tag{2}$$

Thus, if we drop the requirement that the approximating polytope is contained in B_2^n the estimate improves at least by a factor n. We believe that we are able to prove

$$\mathrm{vol}_n(B_2^n \triangle P) \leq \frac{c}{n} \ \mathrm{vol}_n(B_2^n) N^{-\frac{2}{n-1}}$$

but this requires some additional work.

We outline the strategy of the proof. The approximating polytope is obtained in a probabilistic way. We are considering a Euclidean ball that is slightly bigger than the Euclidean ball with radius 1. The factor by which the bigger ball is bigger than the smaller is important and carefully chosen. We are choosing N points randomly on the bigger ball and we are taking the convex hull of these points. With high probability there is a random polytope that fits our requirements.

For technical reasons we are choosing random points on a Euclidean ball of radius 1 and we are approximating a slightly smaller Euclidean ball, say with radius $1 - c_{n,N}$ where $c_{n,N}$ depends on n and N only. In fact, there are two constants a and b such that

$$aN^{-\frac{2}{n-1}} \leq c_{n,N} \leq bN^{-\frac{2}{n-1}} \tag{3}$$

We estimate the expected volume difference between $B_2^n(1 - c_{n,N})$ and a random polytope $[x_1, \ldots, x_N]$ whose vertices are chosen randomly from the boundary of B_2^n

$$\begin{aligned} &\mathbb{E}\mathrm{vol}_n(B_2^n(1 - c_{n,N})\triangle P_N) \qquad (4)\\ &= \int_{\partial B_2^n} \cdots \int_{\partial B_2^n} \mathrm{vol}_n\left(B_2^n(1 - c_{n,N})\triangle[x_1, \ldots, x_N]\right) d\mathbb{P}(x_1) \cdots d\mathbb{P}(x_N). \end{aligned}$$

This gives the result.

Theorem 4. [11] *There is a constant c such that for every $n \in \mathbb{N}$ and for every $M \geq 10^{\frac{n-1}{2}}$ and all polytopes P in $\mathbb{R}^n$ with M facets we have*

$$vol_n(B_2^n \triangle P) \geq c\ vol_n(B_2^n) M^{-\frac{2}{n-1}} \tag{5}$$

Now we are considering general convex bodies.

Theorem 5. *Let K be a convex body in $\mathbb{R}^n$ with C^2-boundary ∂K and everywhere strictly positive curvature κ. Then*

$$\lim_{N \to \infty} \frac{inf\{vol_n(K \setminus P) | P \subseteq K \text{ and } P \text{ has at most } N \text{ vertices}\}}{N^{-\frac{2}{n-1}}} = \frac{1}{2} del_{n-1} \left(\int_{\partial K} \kappa(x)^{\frac{1}{n+1}} d\mu_{\partial K}(x) \right)^{\frac{n+1}{n-1}} \tag{6}$$

where $\mu_{\partial K}$ denotes the surface measure of ∂K.

This theorem gives asymptotically the order of best approximation of a convex body K by polytopes contained in K with a fixed number of vertices. It was proved by McClure and Vitale [15] in dimension 2 and by Gruber [9] for general n. The constant del_{n-1} is positive and depends on the dimension n only. Its order of magnitude can be computed by considering the case $K = B_2^n$. This is the content of Theorem 1 and the refinements found in [13,14]. We get

$$\frac{n-1}{n+1}\mathrm{vol}_{n-1}(B_2^{n-1})^{-\frac{2}{n-1}} \leq \mathrm{del}_{n-1} \leq \left(1+\frac{c\ln n}{n}\right)\frac{n-1}{n+1}\mathrm{vol}_{n-1}(B_2^{n-1})^{-\frac{2}{n-1}}, \tag{7}$$

where c is a numerical constant. In particular,

$$\lim_{n\to\infty}\frac{\mathrm{del}_{n-1}}{n} = \frac{1}{2\pi e}.$$

What happens if we drop the condition that the polytopes have to be contained in the convex body and allow all polytopes having at most N vertices? How much better can we approximate the Euclidean ball?

Theorem 6. [10] *Let K be a convex body in $\mathbb{R}^n$ with C^2-boundary ∂K and everywhere strictly positive curvature κ. Then*

$$\lim_{N\to\infty}\frac{\inf\{vol_n(K\triangle P)|P \text{ is a polytope with at most } N \text{ vertices}\}}{N^{-\frac{2}{n-1}}}$$
$$= \frac{1}{2}ldel_{n-1}\left(\int_{\partial K}\kappa(x)^{\frac{1}{n+1}}d\mu\right)^{\frac{n+1}{n-1}} N^{-\frac{2}{n-1}}.$$

The constant $ldel_{n-1}$ is positive and depends only on n.

By (2) and Theorem 3 there are two positive constants c_1 and c_2 such that for all $n \in \mathbb{N}$

$$\frac{c_1}{n} \leq \mathrm{ldel}_{n-1} \leq c_2.$$

Theorem 7. [9] *Let K be a convex body in $\mathbb{R}^n$ with C^2-boundary ∂K and everywhere strictly positive curvature κ. Then*

$$\lim_{N\to\infty}\frac{\inf\{vol_n(K\triangle P)|K\subseteq P \text{ and } P \text{ is a polytope with at most } N \text{ facets}\}}{N^{-\frac{2}{n-1}}}$$
$$= \frac{1}{2}div_{n-1}\left(\int_{\partial K}\kappa(x)^{\frac{1}{n+1}}d\mu\right)^{\frac{n+1}{n-1}} N^{-\frac{2}{n-1}}.$$

where div_{n-1} is a positive constant that depends on n only.

It is easy to show [10] that there are numerical constants a and b such that $a \cdot n \leq \mathrm{div}_{n-1} \leq b \cdot n$.

Theorem 8. [10] *Let K be a convex body in $\mathbb{R}^n$ with C^2-boundary ∂K and everywhere strictly positive curvature κ. Then*

$$\lim_{N\to\infty} \frac{\inf\{vol_n(K \triangle P)|\ P \text{ is a polytope with at most } N \text{ facets}\}}{N^{-\frac{2}{n-1}}}$$
$$= \frac{1}{2} ldiv_{n-1} \left(\int_{\partial K} \kappa(x)^{\frac{1}{n+1}} d\mu \right)^{\frac{n+1}{n-1}} N^{-\frac{2}{n-1}}.$$

where $ldiv_{n-1}$ is a positive constant that depends on n only.

Clearly, $\mathrm{ldiv}_{n-1} \leq \mathrm{div}_{n-1} \leq c_2 \cdot n$. On the other hand, by Theorem 4 there is a constant c_1 such that for all $n \in \mathbb{N}$

$$\frac{c_1}{n} \leq \mathrm{ldiv}_{n-1}.$$

Now we compare best and random approximation of a convex body by polytopes. As we have already seen in the case of the Euclidean ball that best and random approximation differ by very little. The same holds essentially for arbitrary convex bodies. The probability measure on the boundary of K according to which we are choosing the points has to reflect the surface structure of ∂K. It turns out that the optimal measure is the normalized affine surface area measure.

Theorem 9. [17,18] *Let K be a convex body in $\mathbb{R}^n$ such that there are r and R in $\mathbb{R}$ with $0 < r \leq R < \infty$ so that we have for all $x \in \partial K$*

$$B_2^n(x - rN_{\partial K}(x), r) \subseteq K \subseteq B_2^n(x - RN_{\partial K}(x), R)$$

and let $f : \partial K \to \mathbb{R}_+$ be a continuous, positive function with $\int_{\partial K} f(x) d\mu_{\partial K}(x) = 1$ where $\mu_{\partial K}$ is the surface measure on ∂K. Let κ be the (generalized) Gauß-Kronecker curvature and $\mathbb{E}(f, N)$ the expected volume of the convex hull of N points chosen randomly on ∂K with respect to $\mathbb{P}_f$. Then

$$\lim_{N\to\infty} \frac{vol_n(K) - \mathbb{E}(f, N)}{\left(\frac{1}{N}\right)^{\frac{2}{n-1}}} = c_n \int_{\partial K} \frac{\kappa(x)^{\frac{1}{n-1}}}{f(x)^{\frac{2}{n-1}}} d\mu_{\partial K}(x),$$

where

$$c_n = \frac{(n-1)^{\frac{n+1}{n-1}}\Gamma\left(n+1+\frac{2}{n-1}\right)}{2(n+1)!(vol_{n-2}(\partial B_2^{n-1}))^{\frac{2}{n-1}}}.$$

The minimum at the right-hand side is attained for the normalized affine surface area measure with density

$$f_{as}(x) = \frac{\kappa(x)^{\frac{1}{n+1}}}{\int_{\partial K}\kappa(x)^{\frac{1}{n+1}}d\mu_{\partial K}(x)}.$$

It is clear from Theorem 9 that we get the best random approximation if we choose the affine surface area measure. Then asymptotically the order of magnitude for random approximation is

$$\frac{(n-1)^{\frac{n+1}{n-1}}\Gamma\left(n+1+\frac{2}{n-1}\right)}{2(n+1)!(\mathrm{vol}_{n-2}(\partial B_2^{n-1}))^{\frac{2}{n-1}}}\left(\int_{\partial K}\kappa(x)^{\frac{1}{n+1}}d\mu_{\partial K}(x)\right)^{\frac{n+1}{n-1}}\left(\frac{1}{N}\right)^{\frac{2}{n-1}}$$

and by (6) for best approximation

$$\frac{1}{2}\mathrm{del}_{n-1}\left(\int_{\partial K}\kappa(x)^{\frac{1}{n+1}}d\mu_{\partial K}(x)\right)^{\frac{n+1}{n-1}}\left(\frac{1}{N}\right)^{\frac{2}{n-1}}.$$

Obviously, best approximation is better than random approximation. On the other hand, by the above estimates

$$\left(1-c\frac{\ln n}{n}\right)\lim_{N\to\infty}\frac{\mathrm{vol}_n(K)-\mathbb{E}(f_{as},N)}{\left(\frac{1}{N}\right)^{\frac{2}{n-1}}}$$
$$\leq \lim_{N\to\infty} N^{\frac{2}{n-1}}\inf\{d_S(K,P_N)|P_N\subset K \text{ and } P_N \text{ is a polytope with at most N vertices}\}.$$

In order to verify this we have to estimate the quotient

$$\frac{(n-1)^{\frac{n+1}{n-1}}\Gamma\left(n+1+\frac{2}{n-1}\right)}{2(n+1)!(\mathrm{vol}_{n-2}(\partial B_2^{n-1}))^{\frac{2}{n-1}}}\left(\frac{1}{2}\mathrm{del}_{n-1}\right)^{-1}.$$

By (7) we have $\frac{n-1}{n+1}\mathrm{vol}_{n-1}(B_2^{n-1})^{-\frac{2}{n-1}}\leq \mathrm{del}_{n-1}$. Therefore the quotient is less than $\frac{1}{n!}\Gamma(n+1+\frac{2}{n-1})$. Now we use Stirlings formula to get

$$\frac{\Gamma(n+1+\frac{2}{n-1})}{n!}\leq 1+c\frac{\ln n}{n}.$$

References

1. Aleksandrov A.D. (1939): Almost everywhere existence of the second differential of a convex function and some properties of convex surfaces connected with it. Uchenye Zapiski Leningrad Gos. Univ., Math. Ser., **6**, 3–35.
2. Bangert V. (1979): Analytische Eigenschaften konvexer Funktionen auf Riemannschen Mannigfaltigkeiten. Journal für die Reine und Angewandte Mathematik, **307**, 309–324.
3. Bianchi G., Colesanti A., Pucci C. (1996): On the second differentiability of convex surfaces. Geometriae Dedicata, **60**, 39–48.
4. Böröczky K.(2000): Polytopal approximation bounding the number of k-faces, Journal of Approximation Theory **102**, 263-285.
5. Bronshteyn E.M., Ivanov L.D. (1975): The approximation of convex sets by polyhedra. Siberian Math. J., **16**, 852–853.
6. Evans L.C., Gariepy R.F. (1992): Measure Theory and Fine Properties of Functions. CRC Press.
7. Gordon Y., Reisner S., Schütt C. (1997): Umbrellas and polytopal approximation of the Euclidean ball, Journal of Approximation Theory **90**, 9–22.
8. Gordon Y., Reisner S., Schütt C. (1998): Erratum. Journal of Approximation Theory **95**, 331.
9. Gruber P.M. (1993): Asymptotic estimates for best and stepwise approximation of convex bodies II. Forum Mathematicum **5**, 521–538.
10. M. Ludwig (1999): Asymptotic approximation of smooth convex bodies by general polytopes, Mathematika **46**, 103-125.
11. M. Ludwig, C. Schütt, and E. Werner (2004): Approximation of the Euclidean ball by polytopes, preprint.
12. Macbeath A. M. (1951): An extremal property of the hypersphere. Proc. Cambridge Philos. Soc., **47**, 245–247.
13. Mankiewicz P., Schütt C. (2000): A simple proof of an estimate for the approximation of the Euclidean ball and the Delone triangulation numbers. Journal of Approximation Theory, **107**, 268–280.
14. Mankiewicz P., Schütt C. (2001): On the Delone triangulations numbers. Journal of Approximation Theory, **111**, 139–142.
15. McClure D.E., Vitale R. (1975): Polygonal approximation of plane convex bodies. J. Math. Anal. Appl., **51**, 326–358.
16. Müller J.S. (1990): Approximation of the ball by random polytopes. Journal of Approximation Theory, **63**, 198–209
17. Schütt C., Werner E. (2000): Random polytopes with vertices on the boundary of a convex body, Comptes Rendus de l'Académie des Sciences Paris **331**, 697-201.
18. Schütt C., Werner E. (2003): Polytopes with vertices chosen randomly from the boundary of a convex body, *Israel Seminar 2001-2002*, Lecture Notes in Mathematics 1807 (V.D. Milman and G. Schechtman, eds.), Springer-Verlag, 241–422.

SOME GENERALIZED MAXIMUM PRINCIPLES AND THEIR APPLICATIONS TO CHERN TYPE PROBLEMS

YOUNG JIN SUH*
Department of Mathematics
Kyungpook National University
Taegu, 702-701, KOREA
E-mail: yjsuh@mail.knu.ac.kr

1. Introduction

Before going to give our main talk let us explain an easy global property in Calculus. We consider a function f with one variable on the open interval (a, b). As is well known, if it satisfies

$$f''(x) = 0$$

on (a, b) and if it has a maximum on (a, b), namely, if there is a point x_0 on (a, b) at which $f(x_0) \geq f(x)$ for any point x on (a, b), then f is constant. This property is usually called *Maximum Principle* in Calculus.

Now let us denote by U an open connected set in an m-dimensional Euclidean space $\mathbb{R}^m$ and $\{x^j\}$ a Euclidean coordinate. We denote by L a differential operator defined by

$$L = \sum a^{ij} \frac{\partial^2}{\partial x^i \partial x^j} + \sum b^j \frac{\partial}{\partial x^j},$$

where a^{ij} and b^j are smooth functions on U for any indices. When the matrix (a^{ij}) is positive definite and symmetric, it is called *a second order elliptic differential operator.* We assume that L is an elliptic differential operator. The Maximum Principle is explained as follows:

Maximum Principle. *For a smooth function f on U if it satisfies*

$$Lf \geq 0$$

*This work was supported by grant Proj. No. R14-2002-003-01001-0 from the Korea Research Foundation, Korea 2005.

and if there exists a point in U at which it attains the maximum, namely, if there exists a point x_0 in U at which $f(x_0) \geq f(x)$ for any point x in M, then the function f is constant.

In Riemannian Geometry, this property is reformed as follows. Let (M, g) be a Riemannian manifold with Riemannian metric g. Then we denote by $\triangle$ the Laplacian associated with the Riemannian metric g. A function f is said to be *subharmonic* or *harmonic* if it satisfies

$$\triangle f \geq 0 \quad \text{or} \quad \triangle f = 0.$$

The maximum principle on Riemannian manifolds is as follows:

Maximum Principle. (**[Hopf]**) *For a subharmoni c function f on a Riemannian manifold M if there exists a point in M at which it attains the maximum, then the function f is constant.*

In other words, we have another Maximum Principle different from the above ones:

Maximum Principle. (**[Bochner]**) *On a compact Riemannian manifold M a subharmonic function f on M is constant.*

This property is to give a certain condition for a subharmonic function to be constant. When we give attention to this kind of Maximum Principle, we are able to see the theorem of classical Liouville type.

Liouville's Theorem.

(1) Let f be a subharmonic function defined on $\mathbb{R}^2$. If it is bounded, then it is constant.

(2) Let f be a harmonic function on $\mathbb{R}^m (m \geq 3)$. If it is bounded, then it is constant.

As is already stated, each of these Maximum Principles plays an important role in each branch of Mathematics. Actually *Generalized Maximum Principles* which are later introduced are also important properties to Maximum Principle in a compact Riemannian manifold or more important ones than them.

The purpose of this paper is to prove the fact that

$$\triangle f \geq k f^n \Longrightarrow f = 0, \quad n \geq 1.$$

In order to solve *this kind of Liouville type problem*, we want to investigate all of situations for *any positive real number n* not less than 1. For this problem we want to arrange all the results concerned with this fact.

In section 3 let us show that all the situation greater than 2 could be arrived at the case of Nishikawa's Theorem. By using a new method due to Omori and Yau's maximum principle, we give another proof for the case $n = 2$.

In section 4 we prove another Liouville type theorem for $1 < n < 2$ by using some generalized maximal principles. In section 5 we give some applications of this kind of Liouville type inequality to study complete spacelike hypersurfaces in a Lorentz manifold and S.S. Chern type problems in indefinite complex hyperbolic space.

In section 6 we treat for the case $n = 1$, that is, $\triangle f \geq kf$ for a function f which is bounded from above. Then in this case we can prove that the function f vanishes identically. Moreover, we show a counter example for this type. Namely, there is a smooth unbounded function f which satisfies the above inequality for $n = 1$ but not vanishing. Finally in section 7 we will explain further generalized maximum principles due to Karp [13] and Yau [24].

2. Preliminaries

First of all, let us introduce a *Generalized Maximum Principle* due to Omori [17] and Yau [24]. This is slightly different from the original one.

Theorem 2.1. **([Omori and Yau])** *Let M be an n-dimensional Riemannian manifold whose Ricci curvature is bounded from below on M. Let G be a C^2-function bounded from below on M, then for any $\epsilon > 0$ there exists a point p such that*

$$|\nabla G(p)| < \epsilon, \quad \triangle G(p) > -\epsilon \quad \text{and} \quad \inf G + \epsilon > G(p). \tag{2.1}$$

3. A Liouville Type Theorem for $n \geq 2$

When we consider the generalization of Maximum Principles on a complete Riemannian manifold M, we have two different viewpoints. One is to assume the curvature condition on M and the other is to give the additional condition for the certain function f without the assumption concerning the curvature of M. First of all, we shall consider about the simplest form as follows:

Theorem 3.1. (Generalized Maximum Principle 1. [Nishikawa]) *Let M be a complete Riemannian manifold whose Ricci curvature is bounded from below. If a C^2-nonnegative function f satisfies*

$$\triangle f \geq 2f^2, \tag{3.1}$$

where $\triangle$ denotes the Laplacian on M, then f vanishes identically.

Remark 1. Now suppose that a positive function f satisfies that

$$\triangle f \geq c_0 f^n \tag{3.2}$$

for any real number $n(\geq 2)$. Then we can directly yield

$$\nabla f^{n-1} = (n-1) f^{n-2} \nabla f.$$

So naturally it follows that

$$\begin{aligned}\triangle f^{n-1} &= (n-1)(n-2) f^{n-3} \nabla f \nabla f + (n-1) f^{n-2} \triangle f \\ &\geq (n-1) f^{n-2} \triangle f \\ &\geq c_0 (n-1) f^{2(n-1)}.\end{aligned}$$

We define a function h by f^{n-1}. If $n \geq 2$, then it satisfies

$$\triangle h \geq (n-1) c_0 h^2. \tag{3.3}$$

Thus concerning the Theorem for the case where $n \geq 2$, the condition (3.3) is equivalent to the following

$$\triangle f \geq c_1 f^2,$$

where c_1 is a positive constant.

Remark 2. In the proof of Theorem 3.1, the condition $c_1 = 2$ is essential.

4. A Liouville Type Theorem for $1 < n < 2$

In this section we are going to prove the following

Theorem 4.1. (Generalized Maximum Principal 2) *Let M be a complete Riemannian manifold whose Ricci curvature is bounded from below. If a C^2-nonnegative function f satisfies*

$$\triangle f \geq c_0 f^n, \tag{4.1}$$

where c_0 is any positive constant and n is a real number greater than 1, then f vanishes identically.

Remark 4.1. Concerning the Theorem for the case where $n \geq 2$, the condition (4.1) is equivalent to the following

$$\triangle f \geq c_1 f^2, \tag{4.2}$$

where c_1 is a positive constant. Namely, **Theorem 4.1 is only essential in the case** $1 < n < 2$.

Remark 4.2. In the proof of Theorem 3.1 due to Nishikawa [16], the condition $n = 2$ is essential.

Remark 4.3. When $n = 1$ and the function f is bounded from above, the present author [21] have proved that the Theorem holds true. But until now without any assumption on the function f we consider whether our Theorem is satisfied or not in the case $n = 1$.

Now in order to give a complete proof of Theorem 4.1 we should verify the following Theorem 4.2. That is, we will show another type of Liouville's theorem for $1 < n < 2$ due to Choi, Kwon, Yang and the present author (see [11],[20] and [23]).

Theorem 4.2. *Let M be a complete Riemannian manifold whose Ricci curvature is bounded from below. Let F be any formula of the variable x with constant coefficients such that*

$$F(x) = c_0 x^{n_0} + c_1 x^{n_1} + \cdots + c_k x^{n_k} + c_{k+1}, \tag{4.3}$$

where $n_0 > 1$, $n_0 > n_1 > \cdots > n_k$, $c_0 > 0$ and $c_0 > c_{k+1}$. If a C^2-function f satisfies

$$\triangle f \geq F(f),$$

then we have

$$F(f_0) \leq 0,$$

where f_0 denotes the supremum of the given function f.

5. Applications of Liouville type Theorem 4.2 for the case $1 < n < 2$

Then we will show its applications of Theorem 4.2 to some geometric problems given in [1],[3],[5],[6],[8],[10],[11], [14], [21], [22] and [23]. In order to do this let us introduce the following.

1) Complete space-like hypersurfaces in a Lorentz manifold

Let M' be an $(n+1)$-dimensional Lorentz manifold and let M be a space-like hypersurface of M'. For a point x in M let $\{e_0, e_1, ..., e_n\}$ be a local field of orthonormal frames of M' around of x in such a way that, restricted to M, the vectors $e_1, ..., e_n$ are tangent to M and the other is normal to M. Accordingly, $e_1, ..., e_n$ are space-like vectors and e_0 is a time-like one. For linearly independent vectors u and v in the tangent space T_xM' by which the non-degenerate plane section is spanned, we denote by $K'(u,v)$ the sectional curvature of the plane section in M' and by R' or $Ric'(u,u)$ the Riemannian curvature tensor on M or the Ricci curvature in the direction of u in M', respectively.

Let us denote by ∇' the Riemannian connection on M'. We assume that the ambient space M' satisfies the following conditions; For some constants c_1, c_2 and c_3

$$K'(u,v) \leq -\frac{c_1}{n} \tag{5.1}$$

for any space-like vector u and any time-like vector v,

$$K'(u,v) \geq c_2 \tag{5.2}$$

for any space-like vectors u and v,

$$|\nabla' R'| \leq \frac{c_3}{n}. \tag{5.3}$$

When M' satisfies the above conditions (5.1), (5.2) and (5.3), it is said simply for M' to satisfy the (*) condition.

Remark 5.1. It can be easily seen that if $c_3 = 0$, then the ambient space M' is locally symmetric.

Remark 5.2. Let $\{e_0, e_1, \cdots, e_n\} = \{e_0, e_i\}$, where e_0 and e_i denotes timelike and spacelike vector respectively. Now let us put

$$R'(e_A, e_B)e_C = \sum_D R'_{ABCD} e_D.$$

Then the sectional curvature $K'(e_A, e_B)$ determined by the plane section $[e_A, e_B]$ is given by

$$K'(e_A, e_B) = \frac{g(R'(e_A, e_B)e_B, e_A)}{Q(e_A, e_B)} = \epsilon_A \epsilon_B R'_{ABBA}.$$

So naturally we know $K'(e_0, e_j) = -R'_{0jj0}$ and $K(e_i, e_j) = R'_{ijji}$. From this it follows that

$$\begin{aligned}\text{Ric}'\ (e_A, e_A) =& \sum\nolimits_B \epsilon_B g(R'(e_A, e_B)e_B, e_A)\\ =& \sum\nolimits_B \epsilon_B R'_{ABBA}\\ =& \sum\nolimits_B \epsilon_A K'(e_A, e_B).\end{aligned}$$

So we know

$$\text{Ric}'\ (e_0, e_0) = -\sum\nolimits_j K'(e_0, e_j) = \sum\nolimits_j R'_{0jj0}.$$

From such a notation we know that the condition (5.1) imply

(5.1)′ $$\text{Ric}'\ (v, v) \geq c_1$$

for any time-like vector v. In physics this condition has the meaning of the strong energy condition, that is, the *timelike convergence condition* in Hawking and Ellis [12]. *If M' satisfies the conditions (5.1)', (5.2) and (5.3), it is said simply for M' to satisfy the condition (*').*

Remark 5.3. If M' is a Lorentz space form $M_1^{n+1}(c)$ of index 1 and of constant curvature c, then *it satisfies the condition (*')*, where $-\frac{c_1}{n} = c_2 = c$.

Remark 5.4. Kulkrani's Theorem given in B.O'Neill [18] can not be applied to such a situation that $-\frac{c_1}{n} < c_2$.

Example 1. We give some examples of semi-Riemannian manifold (not Lorenzian space form) satisfyng our $(*)$ condition. Among them the first one is

$$H_1^k(-\frac{c_1}{n}) \times M^{n+1-k}(c_2), c_1 \geq 0, c_2 < 0, -\frac{c_1}{n} \leq c_2.$$

Its sectional curvature is given by

$$K'(u_1, u_b) = -\frac{c_1}{n}, \quad K'(u_a, u_b) = -\frac{c_1}{n},$$

$$K'(u_a, u_r) = 0, \quad K'(u_r, u_s) = c_2,$$

where $a, b, ... = 2, ..k$, $r, s, .. = k+1, ..., n+1-k$, and u_1 denotes time-like and u_a and u_r denotes space-like vector respectively. And the second one is

$$R_1^k \times S^{n+1-k}(1).$$

Its sectional curvature is given by

$$K'(u_1, u_a) = 0, \quad K'(u_a, u_b) = 0,$$

$$K'(u_1, u_r) = 0, \quad K'(u_r, u_s) = 1,$$

where $a, b, ... = 2, ..., k$ and $r, s, .. = k+1, ..., n+1-k$. In particular, $R^1_1 \times S^n(c)$ is said to be *Einstein Static Universe.* Of course, this one is not the Lorentzian space form.

Example 2. Let us consider, so called *Robertson-Walker space-time* $M(c, f)$, (see O'Neill [18] page 343–345) where I denotes an open interval of R^1_1 and $f > 0$ a smooth function defined on the interval I. Then a Lorentz manifold M can be constructed in such a way that

$$M = I \times_f M^3(c), \quad c = -1, 0, 1.$$

Thus its sectional curvature is given by

$$K'(v, w) = (\frac{f'}{f})^2 + \frac{c}{f^2}$$

for any space-like vector v and w and

$$K'(e_0, v) = \frac{g(v, v)}{Q(e_0, v)} \frac{f''}{f}$$

for any time-like vector e_0 and any space-like vector v. When $c > 0$ and $f'' = 0$, it satisfies our assumption (*) for $c_1 = 0$, $c_2 \geq \frac{c}{f^2}$. But until now it is not known whether this can be locally symmetric or not (see Corollary 9. page 345 in O'Neill [18]).

Let $M^m_s(c)$ be an m-dimensional connected semi-Riemannian manifold of index $s(\geq 0)$ and of constant curvature c. It is called a semi-definite space form of index s. In particular, $M^m_1(c)$ is called a Lorentz space form.

In connection with the negative settlement of the Bernstein problem by Calabi [3], Cheng and Yau [5] and Chouquet-Bruhat and et. [6] proved the following famous theorem independently.

Theorem A. *Let M be a complete space-like hypersurface in an $(n+1)$-dimensional Lorentz space form $M^{n+1}_1(c)$, $c \geq 0$. If M is maximal, then it is totally geodesic.*

More generally, complete space-like hypersurfaces with constant mean curvature in a Lorentz manifold are investigated by many differential geometers in various view points; for example Akutagawa [1], Cheng and

Nakagawa [4], Li [15] Nishikawa [16] and Ramanathan [19]. Among them, Nishikawa [16] has considered a complete maximal spacelike hypersurface in a locally symmetric Lorentzian manifold M' satisfying the strong energy condition and nonnegative spacelike sectional curvature. Recently, Li [15] generalized such a result by using a certain curvature condition such that $|\nabla' R'| \leq c_3$, where R' denotes the curvature tensor of the ambient manifold M'.

Now we assume that the ambient space M' satisfies the condition (*) and the mean curvature of the hypersurface M is constant. Namely, we assume that the function h is constant. Then the Laplacian of the squared norm h_2 of the second fundamental form α of M is given by

$$\begin{aligned}\triangle h_2 =& - \triangle(\sum_{i,j} h_{ij}h_{ij}) = -2\sum_{i,j,k}(h_{ijk}h_{ij})_k \\ =& - 2\sum_{i,j,k}(h_{ijkk}h_{ij} + h_{ijk}h_{ijk}) \\ =& 2|\nabla\alpha|^2 - 2\sum_{i,j,k} h_{ijkk}h_{ij} \\ =& 2|\nabla\alpha|^2 - 2\sum_{i,j}(\triangle h_{ij})h_{ij},\end{aligned}$$

where $\nabla\alpha$ is the covariant derivative of the second fundamental form α and $|\nabla\alpha|^2$ is the squared norm of $\nabla\alpha$ which is defined by $-\sum_{i,j,k} h_{ijk}h_{ijk}$. Hence, the assumption of constant mean curvature $\sum_k h_{kkj} = 0$ gives

$$\begin{aligned}\triangle h_2 =& 2|\nabla\alpha|^2 - 2\sum_{i,j}[\sum_k(R'_{0kik;j} + R'_{0ijk;k}) \\ &- \sum_k(h_{kk}R'_{0ij0} + h_{ij}R'_{0k0k}) \\ &- \sum_{k,l}(2h_{kl}R'_{lijk} + h_{lj}R'_{lkik} + h_{li}R'_{kjkl}) \\ &+ hh^2_{ij} - h_2h_{ij}]h_{ij}.\end{aligned}$$

Thus we have

$$\begin{aligned}\triangle h_2 =& 2|\nabla\alpha|^2 - 2\sum_{i,j,k} h_{ij}(R'_{0kik;j} + R'_{0ijk;k}) \\ &+ 2(\sum_{i,j} hh_{ij}R'_{0ij0} - h_2\sum_k R'_{0k0k}) \\ &+ 4\sum_{k,l}(h_{ij}h_{kl}R'_{lijk} - h^2_{lj}R'_{lkjk}) \\ &- 2(hh_3 + h_2^2),\end{aligned}$$

where we have denoted by $h^3_{ij} = \sum h_{ir}h^2_{rj}$ and $h_3 = \sum h^3_{ii}$.

From these formulas it follows that

$$\triangle h_2 \leq 4c_3\sqrt{-h_2} + 2(2nc_2 + c_1)(h_2 + \frac{h^2}{n}) - 2n|h|h_2\sqrt{-h_2} - 2h_2^2. \quad (5.4)$$

Now we define a non-negative function f by $f^2 = -h_2$. Then it turns out to be

$$\triangle f^2 \geq 2\big[f^4 - n|h|f^3 + (2nc_2 + c_1)f^2 - 2c_3 f - \frac{h^2}{n}(2nc_2 + c_1)\big]. \quad (5.5)$$

Now as the first application of Theorem 4.2 for $1 < n < 2$ we introduce the following. In [8] the named authors gave a lower bound of the squared norm of the second fundamental form h_2 of complete space-like hypersurfaces M with constant mean curvature in a Lorentz space form $M_1^m(c)$. Of course in such a case the squared norm h_2 is always negative, that is, $h_2 \leq 0$ on M. That is, we have asserted the following:

Theorem 5.1.([8]) *Let M' be an $(n+1)$-dimensional Lorentz manifold which satisfies the condition (*) and let M be a complete space-like hypersurface with constant mean curvature. If M is not maximal and if it satisfies*

$$2nc_2 + c_1 > 0,$$

then there exist a positive constant a_1 depending on c_1, c_2, c_3, h and n such that $h_2 \geq -a_1$.

Proof. Let $\lambda_1, ..., \lambda_n$ be principal curvatures on M. The Ricci tensor S_{ij} is expressed by

$$S_{ij} = \sum_k (R'_{kijk} - h_{ij}h_{kk} + h_{ik}h_{jk}).$$

So we have

$$S_{jj} \geq (n-1)c_2 - h\lambda_j + \lambda_j^2 \geq (n-1)c_2 - \frac{h^2}{4},$$

which yields the Ricci curvature of M is bounded from below. For the function f defined by $f^2 = -h_2$, by (5.5) we have

$$\triangle f^2 \geq F(f^2),$$

where the function $F(x)$ is defined by

$$F(x) = 2\big[x^2 - n|h|x^{\frac{3}{2}} + (2nc_2 + c_1)x^2 - 2c_3 x^{\frac{1}{2}} - \frac{h^2}{n}(2nc_2 + c_1)\big].$$

By comparing with (4.3) in Theorem 4.2, we know that

$$n = n_0 = 2, c_0 = 2, k = 4, c_5 = 0, \text{ and } n_0 > n_1 > \cdots > n_4.$$

Now we are able to apply Theorem 4.2 to the function f^2. Then we obtain

$$F(f_0^2) \leq 0. \tag{5.6}$$

We define the function $y = y(x)$ of the variable x by

$$\begin{aligned} y &= y(x) \\ &= x^4 - n|h|x^3 + (2nc_2 + c_1)x^2 - 2c_3 x - \frac{h^2}{n}(2nc_2 + c_1). \end{aligned}$$

By the assumption $2nc_2 + c_1 > 0$ and the hypersurface is not maximal, the algebraic equation $y(x) = 0$ with constant coefficients has positive roots, because $y(0) < 0$ and it converges to infinity as x tends to infinity. We denote by $\sqrt{a_1}$ $(a_1 > 0)$ the minimal root among the positive roots. So it depends only on the constant coefficients, namely, it depends on c_1, c_2, c_3, h and n, and by definition we see that

$$y|[0, \sqrt{a_1}) > 0.$$

From the above equation together with (5.6) it follows that we have $0 \leq f_0 \leq \sqrt{a_1}$. Since the squared norm $-h_2$ of the second fundamental form is given by $-h_2 = f^2$, we have

$$-inf\ h_2 = f_0^2 \leq a_1.$$

So we get the conclusion. □

In Theorem A a hypersurface M is totally geodesic means $h_2 = 0$, which is included in such a result of Theorem 5.1. As an application of this result we are able to make a generalization of some theorems which are investigated by Cheng and Nakagawa [4], Li [15], and Nishikawa [16] in new different view points.

Now in this talk let us make a further generalization of Theorem 5.1. Without the assumption of $2nc_2 + c_1 > 0$ we give an extension of Theorem 5.1 in such a way that

Theorem 5.2.([23]) *Let M' be an $(n+1)$ - dimensional Lorentz manifold which satisfies the condition (*) and let M be a complete space-like hypersurface with constant mean curvature H satisfying $h^2 = n^2H^2 \leq -h_2$. Then there exist a positive constant a_2 depending on c_1, c_2, c_3, h and n such that $-a_2 \leq h_2 \leq -h^2$, where $a_2 = -inf\ h_2$.*

When M is maximal, the above condition $n^2H^2 \leq -h_2$ imposed on the constant mean curvature H naturally holds. So without such a condition we have our result in Theorem 5.1.

On the other hand, Akutagawa [1] considered a complete space-like hypersurfaces with constant mean curvature H in a Lorentz space form $M_1^{n+1}(c)$, $c > 0$ and proved that it is totally umbilic under some pinching condition on H. The notion of totally umbilic is much more wider than the notion of totally geodesics. From such a view point, in this talk we consider that the ambient space M' is locally symmetric Lorentz manifold and want to extend the results given in Akutagawa [1], Cheng and Nakagawa [4] and Ramanathan [19]. Then some generalized pinching conditions concerned with constant mean curvature H or concerned with the norm of the second fundamental form h_2 of complete space-like hypersurfaces in M' are given as follows:

Theorem 5.3.([23]) *Let M' be an $(n+1)$-dimensional locally symmetric Lorentz manifold which satisfies the condition (*') and let M be a complete space-like hypersurface with constant mean curvature. Then the following properties hold:*

The case (1): If $n = 2$ and if $h^2 \leq 2(4c_2 + c_1)$, then M is totally umbilic. If $n = 2$ and if it is not totally umbilic, then we have

$$h_2 \geq (2nc_2 + c_1) - h^2.$$

The case (2): If it satisfies $nh^2 - 4(n-1)(2nc_2 + c_1) < 0$, then M is totally umbilic.

The case (3): If it satisfies $nh^2 - 4(n-1)(2nc_2 + c_1) \geq 0$, $h^2 < n(2nc_2 + c_1)$ and it satisfies

$$\begin{aligned} \inf h_2 \geq &- \big[nh^2 - 2(n-1)(2nc_2 + c_1) \\ &- (n-2)|h|\{h^2 - 4(n-1)(2nc_2 + c_1)/n\}^{\frac{1}{2}}\big]/2(n-1), \end{aligned}$$

then M is totally umbilic.

The case (4): If it satisfies $nh^2 - 4(n-1)(2nc_2 + c_1) \geq 0$ and $h^2 \geq n(2nc_2 + c_1)$, then we have

$$\begin{aligned} \inf h_2 \geq &- \big[nh^2 - 2(n-1)(2nc_2 + c_1) \\ &+ (n-2)|h|\{h^2 - 4(n-1)(2nc_2 + c_1)/n\}^{\frac{1}{2}}\big]/2(n-1). \end{aligned}$$

The geometric constants c_1 and c_2 appearing in this theorem are just constants and can not be compared to each other. Though in the proof of Theorem 5.1 we have not used a generalized maximum principle due to Choi, Kwon and the present author [11], but we have used such a maximum principle (see Theorem 4.2) in the proof of Theorem 5.3.

2) Chern type problem in indefinite complex hyperbolic space

Up to now many differential geometers have also interested in the *Chern type* problem which naturally arise from the estimation of the norm of the second fundamental form for the certain kind of complex submanifolds as follows:

Problem *Let M be an n-dimensional complete space-like complex submanifold of an $(n+p)$-dimensional complex hyperbolic space $CH_p^{n+p}(c)$ of constant holomorphic curvature c and of index $2p$. Then does there exist a constant h in such a way that if it satisfies $h_2 > h$, then M is totally geodesic ?*

Now in this talk we consider such a *Chern type* problem in space-like complex submanifolds M of an indefinite complex hyperbolic space $CH_p^{n+p}(c)$ and give a *best possible* estimation of $|\alpha|_2 = h_2$ which is the norm of the second fundamental form α on M.

On the other hand, the complex hyperquadric Q^n in $CH_1^{n+1}(c)$, $c < 0$, is known to be Einstein and the squared norm of the second fundamental form α of Q^n has been estimated by the present author in such a way that

$$h_2 = |\alpha|_2 = \frac{1}{2}nc.$$

Now let us make a generalization of this estimation in a submanifold in $CH_p^{n+p}(c)$. Motivated by such a fact, we want to give a *best possible* estimation for the norm of the second fundamental form α of M in $CH_p^{n+p}(c), c < 0$, as follows:

Theorem 5.4.([11]) *Let M be an n-dimensional complete space-like complex submanifold of an $(n+p)$-dimensional indefinite complex hyperbolic space $CH_p^{n+p}(c)$ of constant holomorphic sectional curvature c and of index $2p$ (> 0). Then it satisfies*

$$h_2 \geq \frac{1}{2}npc,$$

where the equality holds if and only if $p = 1$ and M is globally congruent to a complex quadric Q^n in $CH_1^{n+1}(c)$.

Proof. Since the Ricci curvature $S_{j\bar{j}}$ is given by

$$S_{j\bar{j}} = \frac{n+1}{2}c - h_{j\bar{j}}{}^2,$$

and $h_{j\bar{j}}{}^2 = -\sum_{x,k} h_{jk}{}^x \overline{h}_{jk}{}^x \leq 0$. The Ricci curvature is bounded from below by a negative constant $(n+1)c/2$ and the function f defined by $-h_2$

satisfies the Liouville type inequality

$$2p\Delta f \geq 2f^2 + (n-2)pcf - np^2c^2.$$

Accordingly for the polynomial $F(f)$, we see

$$c_0 = 1/p > c_{k+1} = -npc^2/2, n = 2 \text{ and } n-k = 1.$$

Now we are able to apply above Theorem 4.2 to such a function f. Then it turns out to be

$$(2f_1 + npc)(f_1 - pc) \leq 0,$$

where $f_1 = \sup f = -\inf h_2$. Thus we have $0 \leq f_1 \leq -npc/2$, that is, $0 \geq \inf h_2 \geq npc/2$. It completes the conclusion. The equality $h_2 = npc/2$ holds if and only if α is parallel, the normal curvature is zero and $A = \lambda I$.

Since h_2 is negative and the normal curvature N is constant, the codimension p is equal to 1. Moreover, by the Lemma we know that M is Einstein. This completes the proof. □

Remark 5.5. The complex quadric Q^n in $CH_1^{n+1}(c)$ is an n-dimensional complete space-like complex hypersurface of $CH_1^{n+1}(c)$ and the squared norm $|\alpha|_2 = h_2$ of the second fundamental form α is given by $nc/2$. This means that the estimation given in Theorem 2 is best possible.

Moreover, in this talk we introduce the new notion of *normal curvature* derived from the *normal curvature* tensor which can be naturally defined on the Kaehler submanifold of an indefinite complex hyperbolic space $CH_p^{n+p}(c), c < 0$ (see §3).

In order to get the next result we want to use another technic and another method which are different from the method used in the proof of Theorem 2.1. In this time we will estimate the function $f = -h_2$, which is defined on M, in such a way that $\triangle f \leq kF(f)$, where $F(f)$ denotes some polynomial composed by the function f.

Now, by using such a notion of normal curvature and a different method we want to prove another *Chern type* problem as follows:

Theorem 5.5.([11]) *Let M be an n-dimensional complete space-like complex submanifold of an $(n+p)$-dimensional indefinite complex hyperbolic space $CH_p^{n+p}(c)$ of constant holomorphic sectional curvature c and of index $2p$ (> 0). If the normal curvatures on M are bounded from below by a constant a_1, then there exists a constant β depending on n, p, c and a_1 in such a way that if it satisfies $|\alpha|_2 \geq \beta$, then M is totally geodesic.*

6. A Liouville Type Theorem for $n = 1$

In this section we are going to prove that a Liouville type theorem for the case $n = 1$ while the function f defined on complete Riemannian manifold M is non-negative and bounded from above.

Theorem 6.1.([21]) (Generalized Maximum Principle 3) *Let M be a complete Riemannian manifold whose Ricci curvature is bounded from below. If the non-negative function f is bounded from above and satisfies*

$$(*) \qquad \triangle f \geq kf \text{ for a positive constant } \quad k > 0,$$

then $f = 0$.

Proof. For a constant $a > 0$ let us put $F = (f + a)^{-\frac{1}{2}}$ a smooth positive function. Then we are able to apply H. Omori and S.T. Yau's maximal principle given in section 2. For any $\epsilon > 0$ there exist a point p in M such that

$$|\nabla F|(p) < \epsilon, \triangle F(p) > -\epsilon, F(p) < inf F + \epsilon. \tag{6.1}$$

Then it follows from these properties that we have

$$\epsilon(3\epsilon + 2F(p)) > F(p)^4 \triangle f(p) \geq 0. \tag{6.2}$$

Thus for a convergent sequence $\{\epsilon_m\}$ such that $\epsilon_m > 0$ and $\epsilon_m \to 0$ as $m \to \infty$, there is a point sequence $\{p_m\}$ so that the sequence $\{F(p_m)\}$ satisfies (6.1) and converges to F_0, by taking a subsequence, if necessary, because the sequence $\{F(p_m)\}$ is bounded. From the definition of the infimum and (6.2) we have $F_0 = inf F$ and hence $f(p_m) \to f_0 = sup f$. It follows from (6.1) that we have

$$\epsilon_m\{3\epsilon_m + 2F(p_m)\} > F(p_m)^4 \triangle f(p_m) \tag{6.3}$$

and the left hand side converges to 0 because the function F is bounded. Thus we get

$$F(p_m)^4 \triangle f(p_m) \to 0 \quad (m \to \infty).$$

As is already seen, the Ricci-curvature is bounded from below i.e., so is any λ_B. Since $r = 2\Sigma_B \lambda_B$ is constant, λ_B is bounded from above. Hence $F = (f + a)^{-\frac{1}{2}}$ is bounded from below by a positive constant. From (6.3) it follows that $\triangle f(p_m) \to 0$ as $m \to \infty$. Then by (*) we have that

$$\triangle f(p_m) \geq kf(p_m) \geq 0.$$

Thus we have $f(p_m) \to 0 = inf f$. Since $f(p_m) \to sup f$, $sup f = inf f = 0$. Hence $f = 0$ on M. This completes the above proof of Theorem 6.1. □

Remark 6.1. Let M be a complete n-dimensional Kaehler manifold with constant scalar curvature r. Assume that the totally real bisectional curvature, which can be denoted by its curvature tensor $R_{\bar{i}ij\bar{j}}$, is lower bounded from $b, b > 0$. Then in [14] and [21] we have proved that M is Einstein. In this case we have denoted by the function f

$$f = S_2 - \frac{r^2}{4n} = \sum(\lambda_i - \lambda_j)^2,$$

where the Ricci tensor is given by $S_{i\bar{j}} = \lambda_i \delta_{ij}$ and $S_2 = \sum_{i,j} S_{i\bar{j}} S_{j\bar{i}}$. Moreover we have proved that the function f is bounded from above, and finally we derive the inequality such that

$$\triangle S_2 \geq \sum(\lambda_A - \lambda_B)^2 R_{\bar{A}AB\bar{B}}.$$

So it is equivalent to $\triangle f \geq kf$ for some constant $k > 0$.

Remark 6.2. Let M be an $n(\geq 3)$-dimensional complex submanifold in $P_{n+p}(C)$. If there exist a constant b such that $b > \frac{n^3+2n^2+2n-2}{2n(n^2+2n+3)}$ and $R_{\bar{i}ij\bar{j}} \geq b$, then M is globally isometric to a complex projective space $P_n(C)$. In this case we also have found an inequality of Liouville type such that

$$\triangle h_4 \geq B h_4,$$

where $B = \{2(n^2+2n+3)b - (n^2+2n+2-\frac{2}{n})\}/n$ and $h_4 \leq h_2^2$, $h_2 = \sum \mu_i$ is bounded from above $2h_2 \leq n(n+1)(c-2b)$ and $h_4 = \sum \mu_i^2$.

Then naturally, motivated by the Remark concerned with the scalar curvature, we also have investigated the classification problem with bounded totally real bisectional curvature in [14]. That is, we proved the following

Theorem B. *Let M be an $n(\geq 3)$-dimensional complete complex submanifold of $CH_p^{n+p}(c)$, $p > 0$, with totally real bisectional curvature $\geq b$. Then the following holds*

(1) b is smaller than or equal to $\frac{c}{4}$.

(2) If $b = \frac{c}{4}$, then M is a complex space form $CH^n(\frac{c}{2})$, $p \geq \frac{n(n+1)}{2}$.

(3) If $b = \frac{n(n+p+1)c}{2(n+2p)(n+1)}$, then M is a complex space form $CH^n(\frac{c}{2})$, $p = \frac{n(n+1)}{2}$.

On the other hand, it is seen in Ki and the present author [14] that the squared norm

$$h_2 = |\alpha|_2 = -\sum_{i,j} h_{ij}^x \bar{h}_{ij}^x$$

of the second fundamental form α of M in $CH_p^{n+p}(c)$ satisfies

$$0 \geq |\alpha|_2 \geq n(n+1)\frac{c}{4}$$

the latter equality arising only when M is a complex space form of constant holomorphic sectional curvature $\frac{c}{2}$. However, by estimating the Laplacian of h_2, that is, $\triangle h_2$, we have obtained the same result as in Theorem B with bounded scalar curvature or with bounded totally real bisectional curvature, respectively.

Now in this talk let us investigate the above estimations of $h_2 = |\alpha|_2$, that is, a Chern type problem for *space-like* complex submanifolds M in $CH_p^{n+p}(c)$; more explicitly, for this we will estimate the Laplacian of the squared norm h_4, $h_4 = \sum_j \mu_j^2$, where μ_j denotes an eigenvalue of the Hermitian matrix $H = (h^2_{i\bar{j}})$, which is given by $h^2_{i\bar{j}} = -\sum_{x,k} h^x_{ik}\bar{h}^x_{kj}$. Here we are able to give better estimations than (1.1).

Now let us denote by $a(M)$ the infimum of *totally real bisectional curvatures* of M in $CH_p^{n+p}(c)$. Then we assert the following

Theorem 6.2.([22]) *Let M be an $n = 3$ or $n = 4$ dimensional complete space like complex submanifold of an $(n+p)$-dimensional indefinite complex hyperbolic space $CH_p^{n+p}(c)$ of constant holomorphic sectional curvature $c(>0)$ and of index $2p(>0)$. Then there are constants $a = a(n,p,c)$ and $h = h(n,p,a(M),c)$, $c<0$, so that if $a(M) \geq a$ and the squared norm*

$$|\alpha|_2 = -\sum_{x,i,j} h^x_{ij}\bar{h}^x_{ij}$$

of the second fundamental form α of M satisfies $|\alpha|_2 \geq h$, then M is totally geodesic.

Theorem 6.3.([22]) *Let M be an $n \geq 5$-dimensional complete space-like complex submanifold of an $(n+p)$-dimensional indefinite complex hyperbolic space $CH_p^{n+p}(c)$ of constant holomorphc sectional curvature $c(<0)$ and of index $2p(>0)$ and $p \leq \frac{3(n^2-1)}{n^2-4n-2}$. Then there exists a constant $a = a(n,p,c)$ and a negative constant $h = h(n,p,a(M),c)$ so that if $a(M) \geq a$ and $|\alpha|_2 \geq h$, then M is totally geodesic.*

Remark 6.3. As a final remark we want to show that there exists an example of a smooth function f satisfying (*) but not bounded from above. Let us consider a function f defined by $f(x_1, ..., x_k) = cosh(ax_1)$ on $\mathbb{R}^k$ for some positive constant a. Then it can be easily seen that the function f

satisfies

$$\triangle f = a^2 f.$$

Though it naturally satisfies the inequality (*), but the function f can not be bounded from above and can not be vanishing identically. The condition that f is *bounded from above*, which is given in Theorem 6.1, is essential.

7. Further Remarks

In this section the other generalized maximum principle will be explained in detail.

Theorem 7.1. (Generalized Maximum Principle 4 [Yau]) *Let M be a complete Riemannian manifold whose Ricci curvature is non-negative. If f is positive harmonic, then f is constant.*

Now without the assumption of the curvature we also assert the following

Theorem 7.2. (Generalized Maximum Principle 5 [Yau]) *Let M be a complete Riemannian manifold. If the subharmonic function f satisfies*

$$\int_M |f|^p dM < \infty, \ 1 < p < \infty,$$

then the function f is constant.

In case $p = 1$, the Generalized Maximum Principle 5 is false. In order to deal with the L^1-situation, additional assumption on the curvature of M are necessary. Let ρ be the distance function from any fixed point $x_0 \in M$. Then in this direction we assert the following:

Theorem 7.3. (Generalized Maximum Principle 6 [Karp]) *Let M be a complete Riemannian manifold whose Ricci curvature satisfies*

$$(*) \qquad Ric_M(x) \geq -c\{1 + \rho^2(x)\}$$

for any $x \in M$. If $\triangle f \geq 0$ for $f \geq 0, f \in L^1$, then the function f is constant.

If the Ricci curvature Ric_M is bounded from below, then the above condition is satisfied.

8. Weak Generalized Maximum Principle

Lastly we deal with the weak generalized maximum principle for complete Riemannian manifolds. Let M be a complete Riemannian manifold and

$\rho = \rho_M$ be the distance function from any point x_0 in M. Let $B_r(x_0)$ be the closed geodesic ball of radius r and centered at x_0. Then it is well known that the volume of $B_r(x_0)$ is estimated by the Ricci curvature. By the volume comparision theorem due to Bishop and Crittenden we have

Theorem 8.1. *Let M be an m-dimensional Riemannian manifold. If $Ric \geq -(m-1)\sigma^2$, then there exists a constant c such that*

$$vol\ B_r(x_0) \leq c \int_0^r \frac{(sinh \sigma t)^{m-1}}{\sigma^{m-1}} dt.$$

Now let us define r_2 by

$$r_2 = \lim_{r \to \infty} sup \frac{1}{r^2} log \text{ vol } B_r(x_0).$$

This quantity expresses the rate of the volume growth at infinity of the Riemannian manifold. We notice here that in the case where the volume of M is finite, it is easily seen that $r_2 = 0$.

Proposition. ([Karp]) *Let M be a complete Riemannian manifold. If the Ricci curvature satisfies*

$$Ric_M(x) \geq -c\{1 + \rho_M(x)^2\}$$

for some $c > 0$ and at any $x \in M$, then $r_2 < \infty$.

Theorem 8.2. (Weak Generalized Maximum Principle) *Let M be a complete Riemannian manifold. If $r_2 < \infty$, then*

$$inf \triangle f \leq 0$$

for any bounded function f.

Remark. Assume that the Ricci curvature is bounded from below by a constant $-c(c > 0)$. That is, $Ric \geq -c(c > 0)$. Then it is trivial that

$$Ric_M(x) \geq -c\{1 + \rho_M(x)^2\}$$

and hence we have $r_2 < \infty$ by this proposition.

As an application of Weakley Generalized Maximum Principle we introduce the following. Let N be an Hardmard Cartan manifold and M is immersed in N. Let ρ_N denote the distance function defined by $\rho_N(x) = d(x, x_0)$. We put $f = \rho_N^2$. Then f is the smooth function and by the simple calculation we get

$$\triangle_M f = tr_M[(\nabla^N)^2] f + m g_N(H, grad_N f),$$

where H denotes the mean curvature vector of M. Since M is a complete simply connected Riemannian manifold of non-positive curvature, by the Hessian comparision Theorem due to Greene and Wu [5] to the first term and by the Gauss Lemma we have

$$\triangle_M f \geq 2m - 2mH_0 R,$$

where $H_0 = Sup|H|$ and R denotes the radius of the smallest geodesic ball in N that contains M. Accordingly, the Weak Generalized Maximum Principle (Theorem 8.2) means

$$0 \geq 2m - 2mH_0 R.$$

That implies $H_0 R \geq 1$. Thus we can prove the following

Theorem 8.3. (Karp [13]) *Let M be a complete Riemannian manifold and N a Hardmard Cartan manifold. If $r_2 < \infty$, then M can not be isometrically minimally immersed in a bounded set in N.*

The condition $r_2 < \infty$ is reformed as follows. Assume that the scalar curvature r_M satisfies the following condition:

$$r_M(x) \geq -c\{1 + \rho_M(x)^2\}$$

for some positive constant c at any point $x \in M$. By the Gauss equation we have

$$r_M = \sum_{i \neq j} K_N(e_i, e_j) + n^2 H^2 - S,$$

where H denotes the mean curvature of M and S denotes the squared norm of the second fundamental form. Therefore

$$S(x) \leq n^2 H^2(x) - r_M(x) \leq n^2 H_0^2 + c\{1 + \rho_M(x)^2\},$$

where $H_0 = sup|H|$. Again, by the Gauss equation and the Schwarz inequality we get

$$\begin{aligned} K_M(X, Y) =& K_N(X, Y) + g_N(\alpha(X, X), \alpha(Y, Y)) \\ &- g_N(\alpha(X, Y), \alpha(X, Y)) \\ \leq& g_N(\alpha(X, X), \alpha(Y, Y)) - g_N(\alpha(X, Y), \alpha(X, Y)) \\ \leq& |\alpha(X, X)||\alpha(Y, Y)| - |\alpha(X, Y)|^2 \end{aligned}$$

and hence

$$|K_M(X, Y)(x)| < c_1\{1 + \rho_M(x)^2\}$$

for some positive constant c_1. This yields

$$Ric_M(x) > -c_2\{1 + \rho_M(x)^2\}$$

for some positive constant c_2. Therefore, by Proposition ([Karp]) we have

$$r_2 < \infty.$$

So we can prove the following

Theorem 8.4. (Karp [14]) *Let M be a complete Riemannian manifold and N be a Hardmard Cartan manifold. If the scalar curvature r_M of M satisfies*

$$r_M(x) \geq -c\{1 + \rho_M(x)^2\}$$

for some positive constant c at any point $x \in M$, then M can not be isometrically minimally immersed in a bounded set in N.

If M is compact, the scalar curvature r_M is bounded and hence this condition is satisfied. Accordingly, we have the well known

Theorem 8.5. *Let M be a compact Riemannian manifold and N be a Hardmard Cartan manifold. Then M can not be isometrically minimally immersed in a bounded set in N.*

Acknowledgement. The present author would like to express his sincere gratitude to Professors Eric Grinberg, Gaoyong Zhang and Jiazu Zhou for their invitation to the workshop on Integral Geometry and Convexity Related Topics held at Wuhan, China, on October 18–23, 2004.

References

1. K. Akutagawa , *On space-like hypersurfaces with constant mean curvature in the de Sitter space*, Math. Z. 196(1987), 12–19.
2. J.O. Baek, Q.M. Cheng and Y.J. Suh, *Complete space-like hypersurfaces in locally symmetric Lorentz spaces*, J. of Geometry and Physics, 49(2004), 231–247.
3. E. Calabi, *Examples of Bernstein problems for some nonlinear equations*, Proc. Pure and Appl. Math. 15(1970), 223–230.
4. Q.M. Cheng and H. Nakagawa, *Totally umbilic hypersurfaces*, Hiroshima Math. J. 20(1990), 1–10.
5. S.Y. Cheng and S.T. Yau, *Maximal space-like hypersurfaces in the Lorentz-Minkowski spaces*, Ann. of Math. 104(1976), 223–230.
6. Y. Chouque-Bruhat, A.E. Fisher and J.E. Marsdan, *Maximal hypersurfaces and positivity mass*, Proc. of the E. Fermi Summer School of the Italian Physical Society, J. Ehlers ed. North-Holland, 1979.

7. S.M. Choi, J-H. Kwon and Y.J. Suh, *A Liouville type theorem for complete Riemannian manifolds*, Bull. Korean Math. Soc. 35(1998), 301–309.
8. S.M. Choi, S.M. Lyu and Y.J. Suh, *Complete space-like hypersurfaces in a Lorentz manifold*, Math. J. of Toyama, 22(1999), 53–76.
9. Y.S. Choi, J.-H. Kwon and Y.J. Suh, *On semi-symmetric complex hypersurfaces of a semi-definite complex space form*, Rocky Mountain J. Math. 31(2001), 417–435.
10. Y.S. Choi, J.-H. Kwon and Y.J. Suh, *On semi-Ryan complex submanifolds in an indefinite complex space form*, Rocky Mountain J. Math. **31**(2001), 873–897.
11. Y.S. Choi, J.-H. Kwon and Y.J. Suh, *On Chern type problems in spacelike complex submanifolds of an indefinite complex hyperbolic space*, Houston J. of Math. 30(2004), 35–54.
12. S.W. Hawking and G.F.R. Ellis, *The large scale structure of space-time*, Cambridge University Press, Cambridge, London, NewYork, Melbourne, 1973.
13. L. Karp, *Differential inequalities on complete Riemannian manifolds and applications*, Math. Ann. 272(1985), 449–459.
14. U-H. Ki and Y.J. Suh, *On semi-Kaehler manifolds whose totally real bisectional curvature is bounded from below*, J. Korean Math. Soc., 33(1996), 1009–1038.
15. H. Li, *On complete maximal space-like hypersurfaces in a Lorentz manifold*, Soochow J. Math. 23(1997), 79–89.
16. S. Nishikawa, *On maximal spacelike hypersurfaces in a Lorenzian manifolds*, Nagoya Math. J. 95(1984), 117–124.
17. H. Omori, *Isometric immersions of Riemannian manifolds*, J. Math. Soc. Japan, 19(1967), 205–211.
18. B. O'Neill, *Semi-Riemannian Geometry with Applications to Relativity*, Academic Press, New York, London, 1983.
19. J. Ramanathan, *Complete space-like hypersurfaces of constant mean curvature in de Sitter space*, Indiana Univ. Math. J. 36(1987), 349–359.
20. Y.J. Suh, *Some Liouville-type theorems for complete Riemannian manifolds*, Proc. of the Third Int. Workshop on Diff. Geom., Held at Kyungpook Nat. Univ., Edited by Y.J. Suh, 3(1999), 57–81.
21. Y. J. Suh, *Totally real bisectional curvature and Liouville type inequalities*, Math. J. of Toyama, 22(1999), 35–52.
22. Y. J. Suh, *On a Chern-type problem for space-like Kaehler submanifolds*, Glasnik Mate. 37(57), 331–347, 2002.
23. Y.J. Suh, Y.S. Choi and H.Y. Yang, *On space-like hypersurfaces with constant mean curvature in a Lorentzian manifold*, Houston J. Math. 28(2002), 47–70.
24. S. T. Yau, *Harmonic functions on complete Riemannian manifolds*, Comm. Pure and Appl. Math. 28(1975), 201–228.

FLOATING BODIES AND ILLUMINATION BODIES

ELISABETH WERNER *

Department of Mathematics
Case Western Reserve University
Cleveland, Ohio 44106, U. S. A.
and
Université de Lille 1
UFR de Mathématique
59655 Villeneuve d'Ascq, France
E-mail: emw2@po.cwru.edu

We give an overview over certain geometric bodies, the floating body and the illumination body, associated to a convex body. We also introduce an analytic property for convex bodies, the affine surface area. We describe connections between the introduced concepts.

1. Introduction and Notations

We give an overview over certain geometric bodies associated to a convex body and a certain analytic property defined for a convex body. We describe connections between the introduced concepts. We omit most of the proofs as they have been given in the original papers and we refer to the original articles for the proofs.

We use the following notations. m is the Lebesgue measure on $\mathbb{R}^n$. $B_2^n(a,r)$ is the n-dimensional Euclidean ball with radius r centered at a. We put $B_2^n = B_2^n(0,1)$. By $||.||$ we denote the standard Euclidean norm on $\mathbb{R}^n$, by $<\cdot,\cdot>$ the standard inner product on $\mathbb{R}^n$. For two points x and y in $\mathbb{R}^n$ $[x,y] = \{\alpha x + (1-\alpha)y : 0 \le \alpha \le 1\}$ denotes the line segment from x to y.
For a convex body K in $\mathbb{R}^n$, $\overset{\circ}{K}$ is the interior of K and ∂K is the boundary

*partially supported by a NSF Grant, by a Nato Collaborative Linkage Grant and by a NSF Advance Opportunity Grant.

of K. We also write S^{n-1} for ∂B_2^n. For $x \in \partial K$, $N(x)$ is the outer unit normal vector to ∂K in x. It may not be unique.

$H(x, \xi)$ is the hyperplane containing the point x and orthogonal to ξ. $H^-(x, \xi)$ is the closed halfspace containing the point $x + \xi$, $H^+(x, \xi)$ the other halfspace.

Now we introduce the geometric bodies associated to a given convex body. Let K be a convex body in $\mathbb{R}^n$ and let $t \in \mathbb{R}$, $t \geq 0$.

The *floating body* K_t of K is the intersection of all halfspaces H^+ whose defining hyperplanes H cut off a set of volume t from K

$$K_t = \bigcap_{\int_{K \cap H^-} dm \leq t} H^+$$

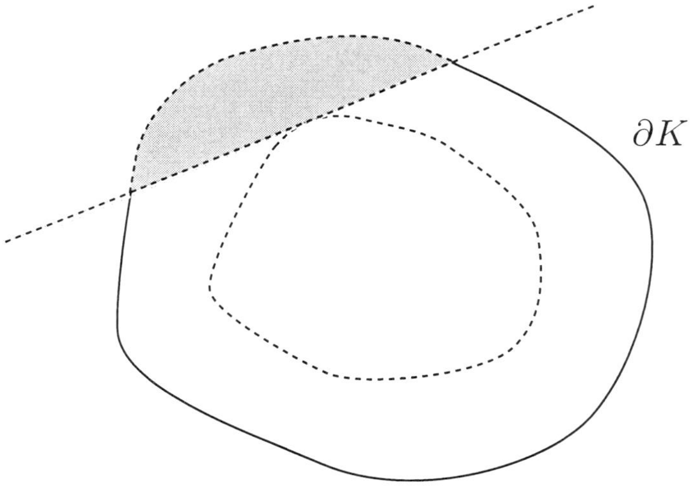

Fig. 1. The hyperplane cuts off a set of volume t.

We choose t small enough so that $K_t \neq \emptyset$.

The *illumination body* K^t of K

$$K^t = \{x \in \mathbb{R}^n : \mathrm{vol}_n\ (co[x, K]) \leq t\}$$

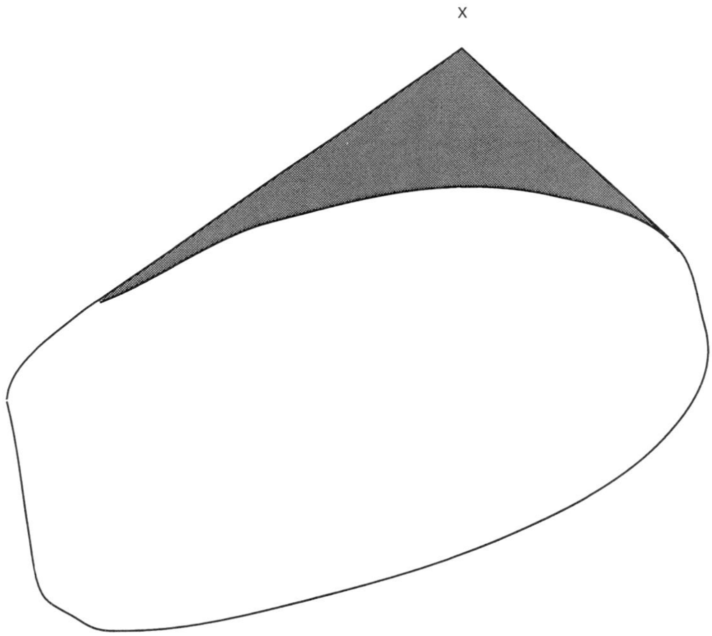

Fig. 2. $\mathrm{vol}_n\ (co[x, K]) \leq t$.

The concept of floating body goes as far back as Dupin and Blaschke [3]. They considered only dimensions 2 and 3. Leichtweiss [9] extended Blaschke's definition to $\mathbb{R}^n$. This definition defines the "boundary" of the floating body $K_{[t]}$ as the set of points that are the center of gravity of all hyperplanes H such that $K \cap H^- = t$. This floating body $K_{[t]}$ is different from the one introduced above. It need not be convex as one can see by considering a triangle in $\mathbb{R}^2$ (see [10]). It is clear that the above floating body K_t is convex as the intersection of convex sets. As far as we know, the the floating body K_t was introduced by Schütt and Werner in [19].

It is obvious that $K_0 = K$ and that $K^0 = K$ and that for all $t \geq 0$ $K_t \subseteq K$ and $K \subseteq K^t$. Further properties are

Proposition 1.

Let K be a convex body in $\mathbb{R}^n$. Then we have

(i) For t small enough, $B_2^n(a,r)_t$ is again a Euclidean ball centered at a with radius

$$r\left(1-\frac{1}{2}\left(\frac{(n+1)\,t}{f_1(\frac{\Delta}{r})\;r^n\;vol_{n-1}(B_2^{n-1})}\right)^{\frac{2}{n+1}}\right),$$

where Δ is the height of a cap of volume δ (see below) and $f_1 : \mathbb{R}^+ \to \mathbb{R}^+$ is a continuous function such that $lim_{s\to 0} f_1(s) = 1$.

(ii) Through every point of ∂K_t there is at least one hyperplane that cuts off a set of volume t from K.

(iii) A supporting hyperplane H of K_t that cuts off a set of volume t touches K_t in exactly one point, namely the barycenter of $K \cap H$.

(iv) K_t is strictly convex.

(v) Let

$$t_0 = \inf\{t : vol_n(K_t) > 0\}$$

Then K_{t_0} consists of one point only.

(vi) K_t is invariant under affine transformations with determinant 1.

Proof.

(i) By symmetry $B_2^n(a,r)_t$ is again a Euclidean ball centered at a. For t small, we now compute the radius ρ_1 of this ball. Let H be a hyperplane that cuts off t of $B_2^n(a,r)$ and let Δ be the height of the cap $B_2^n(a,r) \cap H^-$. Then

$$t = r^n \mathrm{vol}_{n-1}(B_2^n)\int_{1-\frac{\Delta}{r}}^{1}(1-x^2)^{\frac{n-1}{2}}\,dx = \frac{2^{\frac{n+1}{2}}}{n+1} f_1\Big(\frac{\Delta}{r}\Big) r^{\frac{n-1}{2}}\Delta^{\frac{n+1}{2}},$$

where $f_1 : \mathbb{R}^+ \to \mathbb{R}^+$ is a continuous function such that $\lim_{s\to 0} f_1(s) = 1$. Hence

$$\rho_1 = r - \Delta = r\left(1-\frac{1}{2}\left(\frac{(n+1)\,t}{f_1(\frac{\Delta}{r})\;r^n\;\mathrm{vol}_{n-1}(B_2^{n-1})}\right)^{\frac{2}{n+1}}\right).$$

(ii) Let $x \in \partial K_t$. We choose a sequence x_k, $k \in \mathbb{N}$, such that $x_k \notin K_t$ and $\lim_{k\to\infty} x_k = x$. By the definition of K_t we find for every x_k a hyperplane H_k such that $x_k \in H_k^-$ and $\mathrm{vol}_n(K \cap H_k^-) = t$. By compactness there is a subsequence H_{k_j}, $j \in \mathbb{N}$, that converges to a hyperplane H. Therefore $\mathrm{vol}_n(K \cap H^-) = t$ and $x \in H$.

(iii) Let $y \in \partial K_t$ and let $H = H(y,\xi)$ be a supporting hyperplane of K_t that cuts off t from K. The barycenter of $K \cap H$ is computed by

$$z = \frac{1}{\mathrm{vol}_{n-1}(K \cap H)} \int_{K\cap H} x dx$$

Suppose $z \neq y$. We choose

$$\eta = \frac{\xi + \epsilon(y-z)}{\|\xi + \epsilon(y-z)\|}$$

We have

$$\mathrm{vol}_n(K \cap H^-(y,\eta)) =$$

$$\mathrm{vol}_n(K\cap H^-(y,\xi)) + \mathrm{vol}_n(K\cap H^+(y,\xi)\cap H^-(y,\eta)) - \mathrm{vol}_n(K\cap H^+(y,\eta)\cap H^-(y,\xi))$$

We estimate now the volume of the two wedges. Let

$$h(x) = \sup\{s : x - s\xi \in K \cap H^+(y,\xi) \cap H^-(y,\eta)\}$$

Let $\alpha = \alpha(\epsilon)$ be the angle between the two planes. For small α we get that

$$\mathrm{vol}_n(K \cap H^+(y,\xi) \cap H^-(y,\eta)) = \int_{K\cap H(y,\xi)\cap H^-(y,\eta)} h(x)dx + O(\alpha^2).$$

which, for small α, equals

$$\tan\alpha \int_{K\cap H(y,\xi)\cap H^-(y,\eta)} \left|\left\langle \frac{y-z}{\|y-z\|}, x-y \right\rangle\right| dx + O(\alpha^2)$$

The other wedge is handled in the same way and we get that the volume

$$\mathrm{vol}_n(K \cap H^-(y,\eta))$$

for small α equals

$$\mathrm{vol}_n(K \cap H^-(y,\xi)) + \tan\alpha \int_{K\cap H(y,\xi)\cap H^-(y,\eta)} \left|\left\langle \frac{y-z}{\|y-z\|}, x-y \right\rangle\right| dx$$

$$-\tan\alpha \int_{K\cap H(y,\xi)\cap H^+(y,\eta)} \left|\left\langle \frac{y-z}{\|y-z\|}, x-y \right\rangle\right| dx + O(\alpha^2)$$

This equals

$$\mathrm{vol}_n(K \cap H^-(y,\xi)) + \tan\alpha \left\langle \frac{y-z}{\|y-z\|}, \int_{K\cap H(y,\xi)} (x-y)dx \right\rangle + O(\alpha^2) =$$

$$\mathrm{vol}_n(K\cap H^-(y,\xi))+\tan\alpha\ \mathrm{vol}_{n-1}(K\cap H(y,\xi)) \left\langle \frac{y-z}{\|y-z\|}, z-y \right\rangle + O(\alpha^2) =$$

$$\mathrm{vol}_n(K \cap H^-(y,\xi)) - \tan\alpha\ \mathrm{vol}_{n-1}(K \cap H(y,\xi))\|z-y\| + O(\alpha^2)$$

This means that for sufficiently small α this expression is strictly smaller than

$$\mathrm{vol}_n(K \cap H^-(y,\xi))$$

Therefore the hyperplane $H(y,\eta)$ cuts off a set of volume strictly smaller than t. We choose a hyperplane that is parallel to $H(y,\eta)$ and cuts off a set of volume t. But then y is also cut off and cannot be an element of K_t.

(iv) follows from (iii).

□

Proposition 2.

Let K be a convex body in $\mathbb{R}^n$. Then we have

(i) K^t is convex.

(ii) For t small enough, $B_2^n(a,r)^t$ is again a Euclidean ball centered at a with radius

$$\rho_2 = \frac{r}{1-\left(\frac{n(n+1)\ t}{f_2(\frac{r}{\rho_2})\ r^n\ vol_{n-1}(B_2^{n-1})}\right)^{\frac{2}{n+1}}},$$

where $f_2 : \mathbb{R}^+ \to \mathbb{R}^+$ is a continuous function such that $lim_{s\to 1} f_2(s) = 2^{\frac{n+1}{2}}$.

(iii) K^t is invariant under affine transformations with determinant 1.

(iv) For a polytope P in $\mathbb{R}^n$, P^t is again a polytope.

Proof.

(i) To show that the illumination body is convex, it is enough to show that the illumination body of a polytope is convex as every convex body in $\mathbb{R}^n$ can be approximated by polytopes. Let P^t be the illumination body of a polytope P. Let $x, y \in \partial P^t$. Then

$$t = \frac{1}{n} \sum_{\{F:F\ face\ of\ P\}} \mathrm{vol}_{n-1}(F) \max\left\{ < N_F, x - c_F >, 0 \right\}$$

$$= \frac{1}{n} \sum_{\{F:F\ face\ of\ P\}} \mathrm{vol}_{n-1}(F) \max\left\{ < N_F, y - c_F >, 0 \right\},$$

where N_F is the outer normal and c_F is the center of gravity of the face F. Let $0 < \lambda < 1$ and let $z = \lambda\, x + (1-\lambda)\, y$. Then

$$\mathrm{vol}_n([z,P]) \leq \frac{1}{n} \sum_{\{F:F\ face\ of\ P\}} \mathrm{vol}_{n-1}(F) \max\left\{ < N_F, z - c_F >, 0 \right\}$$

$$\leq \frac{\lambda}{n} \sum_{\{F:F\ face\ of\ P\}} \mathrm{vol}_{n-1}(F) \max\left\{ < N_F, x - c_F >, 0 \right\} +$$

$$\frac{1-\lambda}{n} \sum_{\{F:F\ face\ of\ P\}} \mathrm{vol}_{n-1}(F) \max\left\{ < N_F, y - c_F >, 0 \right\} = t.$$

(ii) Again, by symmetry, $B_2^n(a,r)^t$ is a Euclidean ball centered at a. For t small, we now compute the radius ρ_2 of this ball. t equals the volume of a cone minus the volume of a cap. Hence with Proposition 1 (i) we get that

$$t = \mathrm{vol}_{n-1}(B_2^{n-1}) \left(\frac{r^{n-1}\ \rho_2}{n} (1 - \frac{r^2}{\rho_2^2})^{\frac{n+1}{2}} - \frac{2^{\frac{n+1}{2}}\ r^n}{n+1} f_1(1 - \frac{r}{\rho_2})\ (1 - \frac{r}{\rho_2})^{\frac{n+1}{2}} \right)$$

$$= \frac{r^n}{n(n+1)} f_2(\frac{r}{\rho_2})\ (1 - \frac{r}{\rho_2})^{\frac{n+1}{2}}\ \mathrm{vol}_{n-1}(B_2^{n-1}),$$

where $f_2 : \mathbb{R}^+ \to \mathbb{R}^+$ is a continuous function such that $\lim_{s \to 1} f_2(s) = 2^{\frac{n+1}{2}}$. Hence

$$\rho_2 = \frac{r}{1 - \left(\frac{n(n+1)\ t}{f_2(\frac{r}{\rho_2})\ r^n\ \mathrm{vol}_{n-1}(B_2^{n-1})} \right)^{\frac{2}{n+1}}}. \qquad \square$$

Floating bodies and illumination bodies are -in a sense- dual notions. But notice that in general it is not the case that $(K_t)^0 = (K^0)^t$ Indeed, for a polytope P, P^0 is a polytope and hence by Proposition 2 (iv), $(P^0)^t$ is a polytope. However for a polytope P, P_t is strictly convex by Proposition 1 (iv), hence not a polytope.

2. Affine surface area

The affine surface area was originally introduced by Blaschke [3] for convex bodies in $\mathbb{R}^3$ with sufficiently smooth boundary. Its definition involves the Gauss curvature of the boundary points of a convex body. Hence it provides a tool to "measure" the boundary structure of a convex body. Therefore it is not surprising that the affine surface area occurs naturally in problems related to the boundary of a convex body, so for instance in the approximation of convex bodies by polytopes. For more information about this subject and the role the affine surface area plays there, we refer to the works by Bárány, [1,2], Gruber [5,6,7], Schütt [17,18] and Schütt and Werner [20].

Extensions of the affine surface area to higher dimensions and arbitrary convex bodies were only found much later than Blaschke's times by Leichtweiss [9], Lutwak [11,12], Schütt and Werner [19], Schmuckenschläger [16], Meyer and Werner [15] and Werner [24]. Additional references to the affine surface area as well as the proofs of the facts mentioned without their proofs and further applications can also be found in those papers as well as in [14,8,25].

Let K be a convex body in $\mathbb{R}^n$. The affine surface area $as(K)$ is

$$as(K) = \int_{\partial K} \kappa(x)^{\frac{1}{n+1}} d\mu(x),$$

where μ is the surface measure on ∂K, κ the (generalized) Gaussian curvature.

Examples

(i) For every convex polytope P in $\mathbb{R}^n$, $as(P) = 0$. This holds as a.e. on ∂P the Gauss curvature is equal to 0.

(ii) Let $1 \leq p < \infty$ *and let* $B_p^n = \{x \in \mathbb{R}^n : \|x\|_p = (\sum_{i=1}^n |x_i|^p)^{\frac{1}{p}} \leq 1\}$ *be the* l_p^n*-unit ball. Then, for* $1 < p < \infty$*, we have for the affine surface area of* B_p^n

$$as(B_p^n) = \frac{2^n(p-1)^{\frac{n-1}{n+1}}(\Gamma(\frac{p-2}{p(n+1)} + \frac{1}{p}))^{n-1}\Gamma(\frac{p+n-1}{(n+1)p})}{p^{n-1}\Gamma(\frac{p+n-1}{(n+1)p} + (n-1)\frac{1}{p}(\frac{p-2}{n+1} + 1))}$$

For $p = 2$ *we get*

$$as(B_2^n) = 2\frac{\Gamma(\frac{1}{2})^n}{\Gamma(\frac{n}{2})} = n\frac{\pi^{\frac{n}{2}}}{\Gamma(1+\frac{n}{2})} = vol_{n-1}(\partial B_2^n)$$

and for $p = 1$ *and for* $p = \infty$ *we get that*

$$as(B_1^n) = 0 \quad and \quad as(B_\infty^n) = 0,$$

as B_1^n *and* $B_\infty^n = \{x \in \mathbb{R}^n : \|x\|_\infty = max_{1 \leq i \leq n}|x_i| \leq 1\}$ *are polytopes.*

Proof

The Gauss-Kronecker curvature at $x \in \partial B_p^n$ with positive coordinates is

$$\kappa(x) = \frac{(p-1)^{n-1}(\prod_{i=1}^n x_i^{p-2})}{(\sum_{i=1}^n x_i^{2p-2})^{\frac{n+1}{2}}}$$

Thus we get

$$as(B_p^n) = (p-1)^{\frac{n-1}{n+1}} \int_{\partial B_p^n} \frac{(\prod_{i=1}^n |x_i|^{p-2})^{\frac{1}{n+1}}}{(\sum_{i=1}^n |x_i|^{2p-2})^{\frac{1}{2}}} d\mu(x) =$$

$$2^n(p-1)^{\frac{n-1}{n+1}} \int_{B_p^{n-1+}} \left(\prod_{i=1}^{n-1} x_i^{p-2}\right)^{\frac{1}{n+1}} \left(1 - \sum_{i=1}^{n-1} x_i^p\right)^{\frac{n-np-1}{(n+1)p}} d(x_1, \ldots, x_{n-1})$$

By [4] this equals

$$2^n(p-1)^{\frac{n-1}{n+1}} p^{-n+1} \frac{(\Gamma(\frac{p-2}{p(n+1)} + \frac{1}{p}))^{n-1}}{\Gamma(1 - \frac{np+1-n}{(n+1)p} + (n-1)\frac{1}{p}(\frac{p-2}{n+1} + 1))} \Gamma\left(1 - \frac{np+1-n}{(n+1)p}\right) =$$

$$\frac{2^n(p-1)^{\frac{n-1}{n+1}}(\Gamma(\frac{p-2}{p(n+1)} + \frac{1}{p}))^{n-1}\Gamma(\frac{p+n-1}{(n+1)p})}{p^{n-1}\Gamma(\frac{p+n-1}{(n+1)p} + (n-1)\frac{1}{p}(\frac{p-2}{n+1} + 1))}$$

The affine surface area has many nice properties which make it useful for applications. In Proposition 3 we mention some of them. A map T from the set of all convex bodies in $\mathbb{R}^n$ into $\mathbb{R}$ is called a *valuation* if we have for all convex bodies C, K such that $C \cup K$ is again convex

$$T(C \cup K) + T(C \cap K) = T(C) + T(K)$$

Proposition 3. *Let K be a convex body in $\mathbb{R}^n$.*

(i) ***affine invariance***

$$as(T(K)) = T(as(K))$$

for an affine T with det $T = 1$.

(ii) ***valuation***
If C, K, and $C \cup K$ are convex bodies then we have

$$as(C \cup K) + as(C \cap K) = as(C) + as(K)$$

(iii) ***affine isoperimetric inequality***

$$as(K) \leq n vol_n(K)^{\frac{n-1}{n+1}} vol_n(B_2^n)^{\frac{2}{n+1}},$$

with equality iff K is an ellipsoid.

As the affine surface area is related to the boundary structure of of a convex body it comes in naturally in questions of approximation of convex bodies by polytopes:
as(K) "measures" how to distribute the vertices of the approximating polytope on ∂K for optimal approximation (see f.i. [1,2,5,6,7,20]).
Other applications are to characterizations of ellipsoids [21,22], differential equations [13], geometric flows [22], computer vision [23].

Next we want to indicate how the concepts of floating body respectively illumination body and affine surface area are related.

Theorem 4. [19]

Let K be a convex body in $\mathbb{R}^n$. Then

$$lim_{t \to 0} \frac{vol_n(K) - vol_n(K_t)}{t^{\frac{2}{n+1}}} = c_n \int_{\partial K} \kappa(x)^{\frac{1}{n+1}} d\mu(x),$$

where $c_n = \frac{1}{2}\left(\frac{n+1}{vol_{n-1}(B_2^{n-1})}\right)^{\frac{2}{n+1}}$ *is a constant depending only on n.*

and

Theorem 5. [24]

Let K be a convex body in $\mathbb{R}^n$*. Then*

$$lim_{t\to 0}\frac{vol_n(K^t) - vol_n(K)}{t^{\frac{2}{n+1}}} = d_n \int_{\partial K} \kappa(x)^{\frac{1}{n+1}} d\mu(x).$$

where $d_n = \frac{1}{2}\left(\frac{n(n+1)}{vol_{n-1}(B_2^{n-1})}\right)^{\frac{2}{n+1}}$ *is a constant depending only on n.*

Notice that the above two theorems provide a geometric interpretation of the affine surface area in terms of volume differences of a convex body and its floating body respectively of its illumination body. Thus one can use the left hand sides of Theorems 4 and 5 to define the affine surface area for general convex bodies in $\mathbb{R}^n$ and not only for convex bodies with sufficiently smooth boundary as it had been done originally.

References

1. I. Bárány, Random polytopes in smooth convex bodies, Mathematika 39 (1992), 81-92.
2. I. Bárány, Affine perimeter and limit shape, Journal für Reine und Angew. Math.(1997), 71-84.
3. W. Blaschke, Vorlesungen über Differentialgeometrie II: Affine Differentialgeometrie, Springer Verlag, 1923.
4. I.S. Gradsteyn, I.M. Ryzhik, Table of Integrals, Series, and Products, Academic Press, New York, 1965.
5. P.M. Gruber, Approximation of convex bodies, Convexity and its Applications, Birkhäuser, Basel (1983), 131-162.
6. P. Gruber, Aspects of approximation of convex bodies, Handbook of Convex Geometry vol.A, North Holland (1993), 321-345.
7. P. M. Gruber, Asymptotic estimates for best and stepwise approximation of convex bodies II, Forum Math. 5 (1993), 521-538.
8. D. Hug, Contributions to affine surface area, Manuscripta Math. 91 (1996) no. 3, 283-301.
9. K. Leichtweiss, Über ein Formel Blaschkes zur Affinoberfläche, Studia Scient. Math. Hung. 21 (1986), 453-474.
10. K.Leichtweiss, Affine geometry of convex bodies, Johann Ambrosius Barth Verlag, Heidelberg (1998).

11. E. Lutwak, Extended affine surface area, Adv. in Math. 85 (1991), 39-68.
12. E. Lutwak, The Brunn-Minkowski-Firey Theory II: Affine and Geominimal Surface Areas, Adv. in Math. 118 (1996), 244-294.
13. E. Lutwak and V. Oliker, On the regularity of solutions to a generalization of the Minkowski problem, J. Differential Geometry 41 (1995), 227-246.
14. M. Ludwig and M. Reitzner, A characterization of affine surface area, Adv. Math. 147 (1999), no. 1, 138-172.
15. M. Meyer and E. Werner, The Santaló-regions of a convex body, Transactions of the AMS 350, no.11 (1998), 4569-4591.
16. M. Schmuckenschläger, The distribution function of the convolution square of a convex symmetric body in $\mathbf{R}^n$, Israel J. Math. 78 (1992), 309-334.
17. C. Schütt, Random polytopes and affine surface area, Math. Nachrichten 170 (1994), 227-249.
18. C. Schütt, Floating body, Illumination body, and polytopal approximation, Convex geometric analysis (Berkeley, CA, 1996), 203-229, Math. Sci. Res. Inst. Publ., 34, Cambridge Univ. Press, Cambridge (1999).
19. C. Schütt and E. Werner, The convex floating body, Math. Scand. 66 (1990), 275-290.
20. C. Schütt and E. Werner, Polytopes with vertices chosen randomly from the boundary of a convex body, GAFA 2001-2002, Lecture Notes in Mathematics 1807, Springer Verlag, p. 241-422.
21. C. Schütt and E. Werner, Homothetic floating bodies, Geom. Dedic., 49 (1992), 335-348.
22. A. Stancu, The floating body problem, preprint.
23. A. Tannenbaum, Three Snippets of Curve Evolution Theory in Computer Vision, Math.and Computer Modelling Journal 24 (1996), 103-119.
24. E. Werner, Illumination bodies and affine surface area, Studia Math.110 (1994), 257-269.
25. E. Werner, A general geometric construction of affine surface area, Studia Math.132 (3) (1999), 227-238.

APPLICATIONS OF INFORMATION THEORY TO CONVEX GEOMETRY

DEANE YANG *

Department of Mathematics
Polytechnic University
Brooklyn, New York
e-mail: dyang@poly.edu

Connections between geometric invariants of convex bodies and information theoretic invariants of probability densities are described.

1. Introduction

Let X be a finite dimensional real vector space and $dx \in \Lambda^n X^* \backslash \{0\}$. This paper presents some elementary aspects of recent work by Lutwak, Yang, and Zhang [8,32,20,21,22,23,24,25,26,29,28,27,30,31] that extends the L^p Brunn–Minkowski and dual Brunn–Minkowski theories developed by Lutwak [15,16,17,18,19] to establish sharp inequalities satisfied by linear and affine invariants of the following.

- (Geometry) Convex bodies $K \subset X$.
- (Analysis) Functions $f : X \to \mathbf{R}$.
- (Information theory) Probability distributions $f(x)\, dx$ on X.

A question not addressed here and left for the indefinite future is to what extent can the invariants and inequalities be extended to nonlinear spaces? It is not even clear what the correct geometric setting is. All of the results here do not require an inner product or conformal structure. They do, however, rely quite strongly on the flat affine structure of X, leading to manifolds with affine connections as one possible setting for generalizations. Since a fixed convex body plays a role in many of the inequalities, another

*Work supported in part by the National Science Foundation grants dms-014363 and dms-0405707.

direction is to extend the invariants and inequalities to parameterized families of convex bodies, which would include Finsler manifolds.

The invariants for a convex body presented here are unfamiliar to many geometers and analysts, and some may wonder about their significance. Recent work of M. Ludwig [10,11,12,13,14] establish that they are the most fundamental affine and linear geometric invariants of a convex body.

2. Connections between geometry, analysis, and information theory

That information theory is somehow connected to geometric and analytic inequalities is not new. Three examples of this are the following:

- E. Lieb [9] (also, see the work of Cover, Dembo, and Thomas [3]) showed that the Shannon entropy power inequality from information theory and the Brunn–Minkowski inequality from geometry both follow from the sharp Young inequality from analysis proved by Beckner and Brascamp–Lieb.
- W. Beckner and M. Pearson [2] showed that the logarithmic Sobolev inequality of Gross is equivalent to an inequality due to Stam [33] and Weissler [37] involving the entropy and Fisher information of a probability distribution. Stam [33] used this inequality to prove the Shannon entropy power inequality.
- S. Szarek and D. Voiculescu [35,34] introduced the notion of restricted Minkowski sums, established a restricted Brunn–Minkowski inequality, and used it to give a new proof of the Shannon entropy power inequality.

R. Gardner [7] has written an excellent survey article on convex geometric analysis, including its connections to information theory.

3. A little information theory

3.1. *Noisy transmission of a signal*

We begin with a simplified description of a fundamental problem in information theory. A signal is transmitted repeatedly and received with noise. The question is how to use the received noisy signals to obtain the best estimate of what the originally transmitted signal was.

Recall that X is an n-dimensional real vector space. We will always fix a choice of Lebesgue measure dx on X. Let X^* be the dual vector space and $d\xi$ the dual Lebesgue measure.

The transmitted signal is represented by a vector $x_0 \in X$. The received signal is a random vector $x \in X$ with respect to a probability measure

$$p(x - x_0)\, dx,$$

with mean 0. A scalar linear measurement of the received signal is represented by a scalar random variable $\langle \xi, x \rangle$, where $\xi \in X^*$. The mean square error of this measurement is given by

$$\int_X \langle \xi, x - x_0 \rangle^2 p(x - x_0)\, dx = C(\xi, \xi),$$

where $C \in S^2X$ is the *covariance matrix* of the random vector x, given by

$$C = \int_X (x \otimes x) p(x)\, dx.$$

We present two ways of estimating the transmitted signal from repeated transmissions.

3.2. *Mean estimation*

We represent the repeated transmisions by independent identically distributed random vectors $x_1, \ldots, x_N \in X$. The *mean estimate* is given by the random vector

$$\overline{x} = \frac{x_1 + \cdots + x_N}{N};$$

let $p_N(x)\, dx$ be its probability measure. The classical central limit theorem tells us that as the number of transmissions N grows large, the error of the mean estimate goes to zero. In particular, as $N \to \infty$, the mean estimate has the following asymptotic limit:

$$p_N \to G\left(x_0, \frac{C}{N}\right),$$

where $G(x_0, C)$ is the standard normal distribution with mean x_0 and covariance matrix C.

3.3. *Maximum likelihood estimation*

Another estimate is the maximum likelihood estimate, which requires an additional assumption that the probability measure of the noise is known. This approach can be described as follows.

If the transmitted signal is x_0, then the joint distribution of $x_1, \ldots, x_N$ is given by

$$P_N(x_1, \ldots, x_N) = p(x_1 - x_0) \cdots p(x_N - x_0). \qquad (1)$$

Given values for $x_1, \ldots, x_N$, the maximum likelihood estimate is obtained by solve for x_0 that maximizes the right side of (1).

Equivalently, given random vectors $x_1, \ldots, x_N$, let $\tilde{x}$ be the random vector that maximizes the *log likelihood function*,

$$\phi(x) = \sum_{i=1}^{N} \log p(x_i - x).$$

Let q_N be its density function.

Fisher [6] asserted and Doob [5] proved that the error of the maximum likelihood estimate goes to zero and satisfies the following asymptotic estimate: As $N \to \infty$,

$$q_N \to G\left(x_0, \frac{F^{-1}}{\sqrt{N}}\right),$$

where $F \in S^2X^*$ is called the *Fisher information matrix* and is given by

$$F = \int_{\mathbf{R}^n} (d(\log p(x)) \otimes d(\log p(x)))p\, dx. \qquad (2)$$

3.4. *Cramer-Rao inequality*

Observe that C and F are positive definite symmetric matrices. Given two positive definite symmetric matrices A and B, we say $A \geq B$, if $A - B$ is positive semidefinite.

A fundamental theorem in information theory is the Cramer-Rao inequality, which states the following:

Theorem 1. *If C is the covariance matrix and F is the Fisher information matrix of a random vector x, then*

$$C \geq F^{-1}$$

with equality if and only if the random vector is Gaussian.

In other words, the mean square error of the mean estimate is greater than or equal to the mean square error of the maximum likelihood estimate, with equality holding if and only if the distribution is Gaussian.

A fairly trivial consequence is the following.

Corollary 1.

$$(\det C)(\det F) \geq 1.$$

Equality holds if and only if the random vector is Gaussian.

3.5. *Entropy*

The *Shannon entropy* of a random vector $x \in X$ is defined by

$$h[x] = -\int_X p(x) \log p(x)\, dx.$$

Known as Boltzmann entropy in physics. Shannon showed that entropy is related to the amount of information that can be extracted from a noisy signal.

Shannon entropy was extended by Renyi to a more general entropy, given by

$$h_\lambda[x] = \frac{1}{1-\lambda} \log \int_X p(x)^\lambda\, dx,$$

where $\lambda > 0$ is a specified parameter. Many of the mathematical properties of Shannon entropy extend to Renyi entropy, but information theoretic significance of Renyi entropy is still unclear.

We will often use the entropy power, which is defined to be

$$\begin{aligned} N[x] &= \exp h[x] \\ N_\lambda[x] &= \exp h_\lambda[x]. \end{aligned}$$

3.6. *The moment-entropy inequality*

Given an inner product $\langle \cdot, \cdot \rangle$ on X, let g denote the Gaussian random vector on X with covariance matrix $C = \langle \cdot, \cdot \rangle$. Given a random vector x, let $\sigma_2[x]^2$ be the trace of its covariance matrix with respect to the inner product $\langle \cdot, \cdot \rangle$. In other words,

$$\sigma_2[x]^2 = \frac{1}{n} \int \langle x, x \rangle p(x)\, dx.$$

A classical result of information theory (see, for example, Cover and Thomas [4]) is

$$\frac{\sigma_2[x]}{N[x]} \geq \frac{\sigma_2[g]}{N[g]}, \tag{3}$$

with equality holding if and only if $x = tg$ for some $t \in \mathbf{R}$.

If we optimize the left side of (3) over all inner products, we get an affine inequality:

$$\frac{(\det C)^{1/n}}{N[x]} \geq \frac{\sigma_2[g]}{N[g]}, \tag{4}$$

with equality if and only if x is Gaussian.

3.7. *The Fisher information inequality*

Given an inner product $\langle \cdot, \cdot \rangle$ on X, let $\phi_2[x]^2$ be the trace of the Fisher information matrix of a random vector x. In other words,

$$\phi_2[x]^2 = \frac{1}{n} \int \langle d(\log p), d(\log p) \rangle p(x)\, dx.$$

A classical result of information theory (see, for example, Cover and Thomas [4]) is

$$\phi_2[x]N[x] \geq \phi_2[g]N[g], \tag{5}$$

Equality holds if and only if x is Gaussian and has covariance matrix $C = \langle \cdot, \cdot \rangle$.

If we optimize the left side over all inner products, we get an affine inequality:

$$(\det F)^{1/n} N[x] \geq \phi_2[g]N[g], \tag{6}$$

with equality if and only if x is Gaussian.

4. A little convex geometry

A **convex body** is a compact convex set $K \subset X$ that contains the origin in its interior. It is uniquely determined by its **support function** $h_K : X^* \to \mathbf{R}$, where

$$h_K(\xi) = \sup\{\langle \xi, x \rangle \ : \ x \in K\}.$$

It is also uniquely determined by its **dual support function** $h_K^* : X \to \mathbf{R}$, where

$$h_K^*(x) = \inf\{\lambda \ : \ x/\lambda \in K\}.$$

5. Matrices and ellipsoids naturally associated with a convex body

Recall that a positive definite symmetric matrix $A \in S^2X$ defines a quadratic function $A : X \to \mathbf{R}$ and there an ellipsoid $E_A \subset X$, where

$$E_A = \{x \; : \; A^{-1}(x) \leq 1\}.$$

Given an ellipsoid $E \subset X$ centered at the origin, we denote the corresponding matrix by $[E] \in S^2X$.

5.1. *The Legendre ellipsoid*

Associated to any convex body $K \subset X$ is its Legendre ellipsoid $\Gamma_2 K$, which is given by the matrix

$$[\Gamma_2 K] = \frac{1}{V(K)} \int_K x \otimes x \, dx.$$

This is the covariance matrix of the random vector that is uniformly distributed on $K \subset X$.

5.2. *A new ellipsoid*

Lutwak, Yang, Zhang [21] introduced a new ellipsoid $\Gamma_{-2}K$ associated with a convex body K. It is given by the matrix

$$[\Gamma_{-2}K]^{-1} = \frac{1}{V(K)} \int_K \langle dh_K^*(x) \otimes dh_K^*(x) \, dx.$$

Note the resemblance to the Fisher information matrix (2).

6. Geometric inequalities

The information theoretic inequalities presented earlier can be used to prove corresponding geometric inequalities. This is done by defining for each convex body $K \subset X$ a corresponding probability density function

$$p_K(x) = \frac{1}{2^{n/2}\Gamma(\frac{n}{2}+1)V(K)} \exp -h_K^*(x)^2/2.$$

We summarize some results established by Guleryuz, Lutwak, Yang, and Zhang[8].

6.1. *The moment-volume inequality*

Applying (4) to the density p_K yields the classical geometric inequality

$$V(\Gamma_2 K) \geq V(K),$$

with equality if and only if K is an ellipsoid centered at the origin.

6.2. *A dual inequality*

Applying (6) to the density p_K yields the following inequality established by Lutwak, Yang, and Zhang[21]:

$$V(\Gamma_{-2} K) \leq V(K),$$

with equality if and only if K is an ellipsoid centered at the origin.

6.3. *The Cramer-Rao inequality for convex bodies*

Applying Theorem 1 to the density p_K yields the following inclusion established by Lutwak, Yang, and Zhang[24]:

$$\Gamma_{-2} K \subset \Gamma_2 K,$$

with equality if and only if the convex body K is an ellipsoid centered at the origin.

7. Moment-entropy inequalities

In the previous section we showed how information theoretic inequalities could be used to prove geometric inequalities. Lutwak, Yang, Zhang[26] have also used a geometric inequality to prove an information theoretic-inequality.

Lutwak and Zhang [32] proved the following affine geometric inequality:

Theorem 7.1. *Given $n, p \geq 1$, there is an explicit constant $c(n,p) > 0$ such that if $S \subset X$ and $\Sigma \subset X^*$ are star bodies, then*

$$\left(\int_{\Sigma \times S} |\langle \xi, x \rangle|^p \, d\xi \, dx \right)^{\frac{1}{p}} \geq c(n,p)[V(S)V(\Sigma)]^{\frac{1}{p}+\frac{1}{n}}. \tag{7}$$

Equality holds if and only if $S = E$ and $\Sigma = tE^*$ for some $t > 0$ and ellipsoid $E \subset X$ centered at the origin.

Given an ellipsoid $E \subset X$ centered at the origin, let z_E denote the Gaussian random vector with mean 0 and covariance matrix $[E]$. Theorem 7.1 is used to prove the following moment-entropy inequality[26]:

Theorem 7.2. *Given $n \geq 1$, $p \in [1, \infty)$, and $\lambda \in (n/(n+p), \infty]$, there is an explicit constant $c(n,p,\lambda)$ such that if x is a random vector in X with density function f and ξ a random vector in X^* with density function ϕ, both with finite p-th moments, then*

$$E[|\langle \xi, x\rangle|^p] \geq c(n,p,\lambda)(N_\lambda[x]N_\lambda[\xi])^{p/n}.$$

Equality holds if and only if $x = tz_E$ and $\xi = \tau z_{E^*}$ for some centered ellipsoid $E \subset L$ and positive reals t and τ.

8. Affine Sobolev inequalities

Lutwak-Yang-Zhang [25] established a family of L^p affine Sobolev inequalities that generalize both the inequality (6) and the sharp Sobolev inequalities of Aubin [1] and Talenti [36].

References

1. T. Aubin, *Problèmes isopérimétriques et espaces de Sobolev*, J. Differential Geometry **11** (1976), 573–598.
2. William Beckner and Michael Pearson, *On sharp Sobolev embedding and the logarithmic Sobolev inequality*, Bull. London Math. Soc. **30** (1998), 80–84.
3. Thomas M. Cover, Amir Dembo, and Joy A. Thomas, *Information theoretic inequalities*, IEEE Transactions on Information Theory **37** (1991), 1501–1518.
4. Thomas M. Cover and Joy A. Thomas, *Elements of information theory*, John Wiley & Sons Inc., New York, 1991, A Wiley-Interscience Publication.
5. J. L. Doob, *Probability and statistics*, Trans. Amer. Math. Soc. **36** (1934), no. 4, 759–775. MR 1 501 765
6. R. A. Fisher, *Theory of statistical estimation*, Philos. Trans. Roy. Soc. London Ser A **222** (1930), 309–368.
7. R. J. Gardner, *The Brunn-Minkowski inequality*, Bull. Amer. Math. Soc. (N.S.) **39** (2002), 355–405.
8. Onur G. Guleryuz, E. Lutwak, D. Yang, and G. Zhang, *Information-theoretic inequalities for contoured probability distributions*, IEEE Trans. Inform. Theory **48** (2002), 2377–2383.
9. Elliott H. Lieb, *Proof of an entropy conjecture of Wehrl*, Comm. Math. Phys. **62** (1978), no. 1, 35–41. MR 80d:82032
10. M. Ludwig, *Moment vectors of polytopes*, Rend. Circ. Mat. Palermo (2) Suppl. (2002), no. 70, part II, 123–138, IV International Conference in "Stochastic Geometry, Convex Bodies, Empirical Measures & Applications to Engineering Science", Vol. II (Tropea, 2001).

11. ______, *Projection bodies and valuations*, Adv. Math. **172** (2002), 158–168.
12. ______, *Valuations of polytopes containing the origin in their interiors*, Adv. Math. **170** (2002), 239–256.
13. ______, *Ellipsoids and matrix-valued valuations*, Duke Math. J. **119** (2003), 159–188.
14. ______, *Minkowski valuations*, Trans. Amer. Math. Soc. **357** (2005), 4191–4213.
15. E. Lutwak, *Centroid bodies and dual mixed volumes*, Proc. London Math. Soc. **60** (1990), 365–391.
16. ______, *Extended affine surface area*, Adv. Math. **85** (1991), 39–68.
17. ______, *Inequalities for mixed projection bodies*, Trans. Amer. Math. Soc. **339** (1993), 901–916.
18. ______, *The Brunn-Minkowski-Firey theory. I. Mixed volumes and the Minkowski problem*, J. Differential Geom. **38** (1993), 131–150.
19. ______, *The Brunn-Minkowski-Firey theory. II. Affine and geominimal surface a reas.*, Adv. Math. **118** (1996), 244–294.
20. E. Lutwak, D. Yang, and G. Zhang, *L_p affine isoperimetric inequalities*, J. Differential Geom. **56** (2000), 111–132.
21. ______, *A new ellipsoid associated with convex bodies*, Duke Math. J. **104** (2000), 375–390.
22. ______, *A new affine invariant for polytopes and Schneider's projection problem*, Trans. Amer. Math. Soc. **353** (2001), 1767–1779 (electronic).
23. ______, *Sharp L_p Sobolev and logarithmic Sobolev inequalities*, preprint, 2001.
24. ______, *The Cramer-Rao inequality for star bodies*, Duke Math. J. **112** (2002), 59–81.
25. ______, *Sharp affine L_p Sobolev inequalities*, J. Differential Geom. **62** (2002), 17–38.
26. ______, *Moment-entropy inequalities*, Annals of Probability **32** (2004), 757–774.
27. ______, *On the L^p Minkowski problem*, Trans. Amer. Math. Soc. **356** (2004), 4359–4370.
28. ______, *Volume inequalities for subspaces of L_p*, J. Differential Geom. **68** (2004), 159–184.
29. ______, *Cramer-Rao and moment-entropy inequalities for Renyi entropy and generalized Fisher information*, IEEE Trans. Inform. Theory **51** (2005), 473–478.
30. ______, *L^p John ellipsoids*, Proc. London Math. Soc. **90** (2005), 497–520.
31. ______, *Volume inequalities for isotropic measures*, preprint, 2005.
32. E. Lutwak and G. Zhang, *Blaschke-Santalo inequalities*, J. Differential Geom. **47** (1997), 1–16.
33. A. J. Stam, *Some inequalities satisfied by the quantities of information of Fisher and Shannon*, Information and Control **2** (1959), 101–112. MR 21 #7813
34. S. J. Szarek and D. Voiculescu, *Shannon's entropy power inequality via restricted Minkowski sums*, Geometric aspects of functional analysis, Lecture

Notes in Math., vol. 1745, Springer, Berlin, 2000, pp. 257–262.
35. Stanislaw J. Szarek and Dan Voiculescu, *Volumes of restricted Minkowski sums and the free analogue of the entropy power inequality*, Comm. Math. Phys. **178** (1996), no. 3, 563–570.
36. G. Talenti, *Best constant in Sobolev inequality*, Ann. Mat. Pura Appl. (4) **110** (1976), 353–372.
37. Fred B. Weissler, *Logarithmic Sobolev inequalities for the heat-diffusion semigroup*, Trans. Amer. Math. Soc. **237** (1978), 255–269. MR 80b:47057

CONTAINMENT MEASURES IN INTEGRAL GEOMETRY

GAOYONG ZHANG*

Department of Mathematics
Polytechnic University
Brooklyn, NY 11201, USA
E-mail: gzhang@math.poly.edu

JIAZU ZHOU†

College of Mathematics and Computer Sciences
Guizhou Normal University
Guiyang, Guizhou 550001, China
E-mail: jzhou07828@yahoo.com

(Dedicated to Professor Delin Ren on the occasion of his 70th birthday.)

Containment measure is the kinematic measure of a random convex body contained in a fixed convex body. This paper surveys results of containment measures related to integral formulas, isoperimetric inequalities, and geometric probability.

1. Introduction

Kinematic formulas in integral geometry are integral formulas that represent integrals of geometric functionals on the intersection of fixed and moving geometric figures by geometric invariants of the figures themselves (see [17,21,20,8]). These formulas can be viewed as integral formulas for various *intersection measures*. They are useful for solving problems in geometric probability and stochastic geometry. Some problems in geometric probability require more tools other than intersection measures. For instance, solutions to the Buffon needle problem of lattices need to compute the measure of the needle that is contained in a fundamental region of the lattice. The kinematic measure of a moving geometric figure that is contained in a fixed geometric figure is called a *containment measure*. Delin Ren [14] intro-

*Work supported in part by NSF grant DMS-9906856.
†Work supported in part by Guizhou Province Scientific Research Foundation.

duced the notion of generalized support function of a convex body in the plane and used it to establish integral formulas for the containment measure of a line segment inside a convex body. He then applied his formulas to solving generalized Buffon needle problems of lattices (see [13]). One of the aims of this survey article is to present integral formulas of containment measures.

In the plane $\mathbb{R}^2$, kinematic formulas provide different approaches and generalizations to isoperimetric problems (see [17] for details). One of the approaches is to use containment measures. The positivity of containment measures gives sufficient conditions for a geometric figure to be contained in another. Such conditions in terms of areas and perimeters in $\mathbb{R}^2$ were given by Hadwiger (see [17]). Ren [13] obtained similar sufficient conditions in $\mathbb{R}^2$ that have a close connection with the isoperimetric problem. Sufficient conditions for the containment of convex bodies in $\mathbb{R}^n$ were obtained in [26], which imply Bonnesen-type isoperimetric inequalities. Integrals of square curvatures and differential geometric techniques play roles in sufficient conditions for containment of smooth domains in a series of work of Zhou ([27] – [35]). We will give a description of these results.

2. Containment measures of domains in homogeneous spaces

Let G be a Lie group that acts on a C^∞ manifold X, and let μ be an invariant measure on G. A domain in X is the closure of an open set. If K and D are domains in X, the containment measure of K inside D is defined as

$$m_1(K, D) = \int_{K \subseteq gD} d\mu(g).$$

Let k be a positive integer. The k-th containment measure of K in k copies of D is defined as

$$m_k(K, D) = \int_{K \subseteq \bigcup_{i=1}^{k} g_i D} d\mu(g_1) \cdots d\mu(g_i).$$

We consider the invariant measure μ that comes from an invariant density ω on a homogeneous space G/H, where H is a closed subgroup of G. Let us recall the construction of a density ω.

Suppose that the left cosets gH, $g \in G$, are integral manifolds of a completely integrable Pfaffian system

$$\omega_1 = 0, \ \omega_2 = 0, \ \ldots, \ \omega_m = 0,$$

where ω_i are left invariant 1-forms on G. The differential m-form

$$\omega = \omega_1 \wedge \omega_2 \wedge \cdots \wedge \omega_m$$

is left invariant under G. If ω is also right invariant under H, that is, $\omega(gg_1) = \omega(g)$, $g \in G$, $g_1 \in H$, then ω is an invariant differential m-form on the homogeneous space G/H, called a density on G/H. If such an invariant density on G/H exists, then it is unique up to a constant factor. The invariant measure defined by this invariant density is called the Haar measure. A necessary and sufficient condition for the differential m-form ω on G to be a density on G/H is that it is closed, that is, $d\omega = 0$.

2.1. *Containment measures of convex bodies*

Let G_n be the group of rigid motions in $\mathbb{R}^n$. For every rigid motion $g \in G_n$, there are a unique rotation $\alpha \in SO(n)$ and a unique translation $t \in \mathbb{R}^n$ so that $gx = \alpha x + t$. Let $d\alpha$ be the invariant density of $SO(n)$ normalized so that $\int_{SO(n)} d\alpha = 1$, and let dt be the volume element of $\mathbb{R}^n$. The invariant density dg of G_n is given by

$$dg = dt \wedge d\alpha.$$

A convex body in $\mathbb{R}^n$ is a compact convex set with nonempty interior. Let D be a convex body in $\mathbb{R}^n$, and let K be a compact convex set in $\mathbb{R}^n$. The containment measure of K inside D is

$$m_1(K, D) = \int_{\{g \in G_n : K \subseteq gD\}} dg.$$

2.2. *Containment measures of convex bodies in strips*

Let D be a strip bounded by two parallel hyperplanes with distance $b > 0$ in $\mathbb{R}^n$. The isotropy group $H = \{g \in G_n : gD = D\}$ is a closed subgroup and is isomorphic to the group of rigid motions in $\mathbb{R}^{n-1}$. The invariant density $d\nu$ of the homogeneous space G_n/H can be represented as

$$d\nu = dh \wedge du,$$

where h is the distance of the central hyperplane of D from the origin, u is the unit normal of the central hyperplane, and du is the surface area element of the unit sphere S^{n-1}.

3. Integral formulas for containment measures

If K is a convex body in $\mathbb{R}^n$, the support function $h(u)$ of K is defined as a function on the unit sphere S^{n-1} by

$$h(u) = \{u \cdot x : x \in K\}, \quad u \in S^{n-1}.$$

The width function $b(u)$ of K is defined by

$$b(u) = h(u) + h(-u).$$

Denote by $V(K)$ the volume of a convex body K. The quermassintegrale $W_i(K)$ of K are defined as the coefficients of the Steiner polynomial,

$$V(K + rB_n) = \sum_{i=0}^{n} \binom{n}{i} r^i W_i(K),$$

where B_n is the unit ball. Note that $W_0(K) = V(K)$, $W_1(K) = S(K)/n$, and $W_{n-1}(K) = \frac{1}{2}V(B_n)B(K)$, where $S(K)$ is the surface area of K and $B(K)$ is the mean width of K.

Denote by $m(\ell, K)$ the containment measure of a line segment N of length ℓ inside a convex body K, that is, $m(\ell, K) = m_1(N, K)$.

3.1. *Containment measures of line segments inside convex bodies in $\mathbb{R}^2$*

Let L_k, $1 \leq k \leq n-1$, be a k-dimensional plane in $\mathbb{R}^n$. All the set of all k-planes in $\mathbb{R}^n$ is a homogeneous space. It has an invariant density, denoted by dL_k and can be represented as

$$dL_k = dx \wedge d\alpha,$$

where dx is the volume element in the $(n-k)$-dimensional subspace orthogonal to L_k, and $d\alpha$ is the density of rotation group $SO(k)$ in L_k that is normalized so that $\int_{SO(k)} d\alpha = 1$.

Theorem 3.1. ***(Delin Ren)*** *Let $K \subset \mathbb{R}^2$ be a convex body with area F and perimeter L, and let $N \subset \mathbb{R}^2$ be a line segment with length ℓ. The containment measure of N inside K is given by the integral formula*

$$m(\ell, K) = F - \frac{\ell L}{\pi} + \int_{K \cap L_1 \neq \phi} (\ell - \sigma)_+ dL_1,$$

where $\sigma = |K \cap L_1|$ is the chord length K intercepted by the line L_1 in $\mathbb{R}^2$.

Suppose a straight line L_1 in $\mathbb{R}^2$ has equation,

$$x\cos\varphi + y\sin\varphi = h.$$

Delin Ren [14] introduced the notion of generalized support function of a convex body in $\mathbb{R}^2$. He represented the containment measure of a line segment inside a convex body by using the generalized support function. For a convex body K in $\mathbb{R}^2$, the generalized support function $h(\sigma,\varphi)$, $0 \le \sigma < \infty$, $0 \le \varphi \le 2\pi$, is defined as

$$h(\sigma,\varphi) = \begin{cases} \sup_{L_1}\{h : |K \cap L_1| \ge \sigma\}, & \sigma \le \sigma_M(\varphi) \\ 0, & \sigma > \sigma_M(\varphi), \end{cases}$$

where $\sigma_M(\varphi)$ is the maximum of the chord length $|K \cap L_1|$ when φ is fixed. The function $h(0,\varphi)$ is the standard support function of a convex body.

Theorem 3.2. ***(Delin Ren)*** *Let $K \subset \mathbb{R}^2$ be a convex body with generalized support function $h(\sigma,\varphi)$, and let $N \subset \mathbb{R}^2$ be a line segment of length ℓ. The containment measure of N inside K is given by the integral formula*

$$m(\ell,K) = \frac{1}{\pi}\int_0^{2\pi}\int_\ell^\infty h(\sigma,\varphi)\,d\sigma.$$

3.2. *Containment measures of line segments inside convex bodies in $\mathbb{R}^n$*

Let α_n be the surface area of the unit sphere S^{n-1}. Delin Ren proved the following integral formula for the containment measure $m(\ell,K)$. It implies Crofton formulas for both the volume and the surface area of a convex body.

Theorem 3.3. ***(Delin Ren)*** *Let $K \subset \mathbb{R}^n$ be a convex body with volume V and surface area S, and let $N \subset \mathbb{R}^n$ be a line segment with length ℓ. The containment measure of N inside K is given by the integral formula*

$$m(\ell,K) = V - \frac{\alpha_{n-1}\ell S}{(n-1)\alpha_n} + \int_{K\cap L_1 \ne \phi} (\ell-\sigma)_+ dL_1,$$

where $\sigma = |K \cap L_1|$ is the chord length of K intercepted by the line L_1 in $\mathbb{R}^n$.

If the length of the line segment N is greater than the diameter of the convex body K, then N can not be contained in K, and thus, $m(\ell,K) = 0$. By this fact and the theorem above,

Corollary 3.4. *If ℓ is a number greater than the diameter of a convex body K, then*

$$V - \frac{\alpha_{n-1}\ell S}{(n-1)\alpha_n} + \int_{K\cap L_1\neq\phi} (\ell-\sigma)dL_1 = 0.$$

The identity above gives the Crofton formulas,

$$\int_{K\cap L_1\neq\phi} \sigma dL_1 = V, \quad \int_{K\cap L_1\neq\phi} dL_1 = \frac{\alpha_{n-1}S}{(n-1)\alpha_n}.$$

The projection of a convex body K on to the subspace $u^\perp$, $u \in S^{n-1}$, is a convex body in $u^\perp$. The notion of projection was generalized by Ren and Zhang [13]. Let L_1 is a straight line in $\mathbb{R}^n$. For a convex body K, the set

$$K(\sigma, u) = \{u^\perp \cap L_1 : L_1 \parallel u,\ |K \cap L_1| \geq \sigma\}$$

is a convex body in $u^\perp$, which is called the chord projection of K. The set $K(0,u)$ is the standard projection of K. The $(n-1)$-dimensional volume of $K(\sigma,u)$ is called the chord projection function of K, denoted by $A(\sigma,u)$. Let $\sigma_M(u)$ be the maximal chord length of K in direction u. When $\sigma > \sigma_M(u)$, $K(\sigma,u)$ is empty, and $A(\sigma,u)$ is defined as 0.

The containment measure of a line segment inside a convex body can be represented by the chord projection function.

Theorem 3.5. ***(Ren and Zhang)*** *Let $K \subset \mathbb{R}^n$ be a convex body with chord projection function $A(\sigma,u)$, and let $N \subset \mathbb{R}^n$ be a line segment of length ℓ. The containment measure of N inside K is given by the integral formula*

$$m(\ell, K) = \frac{1}{\alpha_n}\int_{S^{n-1}} du \int_\ell^\infty A(\sigma,u)\,d\sigma.$$

3.3. *Containment measures of convex bodies inside parallelotopes*

Let $K \subset \mathbb{R}^n$ be a convex body with width function $b(u)$, and let P be a parallelotope defined by its support function

$$h_P(u) = \frac{1}{2}\sum_{i=1}^n a_i \frac{|u_1 \wedge \cdots \wedge u \wedge \cdots \wedge u_n|}{|u_1 \wedge \cdots \wedge u_i \wedge \cdots \wedge u_n|}$$

for $a_i > 0$ and linearly independent $u_i \in S^{n-1}$, that is, u_i are the unit normals of the facets of P and a_i are the corresponding widths.

Theorem 3.6. ***(Ren and Zhang)*** *If K is a convex body and P is a parallelotope in $\mathbb{R}^n$, then the containment measure of K inside P is given by* [16]

$$m_1(K,P) = |u_1 \wedge \cdots \wedge u_n|^{-1} \int_{SO(n)} \prod_{i=1}^{n} b(a_i, \alpha u_i)\, d\alpha,$$

where $b(a_i, u_i) = (a_i - b(u_i))_+$.

3.4. *Examples*

Rectangle. If K is a rectangle of sides $a \geq b > 0$, then[14]

$$m(\ell, K) = \frac{1}{\pi}\begin{cases} \pi ab - 2(a+b)l + l^2, & \text{if} \quad 0 \leq l \leq b, \\ \pi ab - 2ab \arccos \frac{b}{l} & \\ \quad -2al + 2a(l^2 - b^2)^{1/2} - b^2, & \text{if} \quad b \leq l \leq a, \\ 2a(l^2 - b^2)^{1/2} + 2b(l^2 - a^2)^{1/2} & \\ \quad -a^2 - b^2 - l^2 + 2ab \arcsin \frac{a}{l} - 2ab \arccos \frac{b}{l}, & \text{if} \quad a \leq l \leq d. \end{cases}$$

Triangle. If K is an equilateral triangle of side a, then [11]

$$m(\ell, K) = \frac{1}{\pi}\begin{cases} \frac{\sqrt{3}}{4}\pi a^2 - 3al + \frac{\sqrt{3}}{6}\pi l^2 + \frac{3}{4}l^2, & 0 \leq l \leq \frac{\sqrt{3}}{2}a \\ \frac{\sqrt{3}}{4}\pi a^2 - 3al + \frac{9}{2}a(l^2 - \frac{3}{4}a^2)^{1/2} + \frac{\sqrt{3}}{6}\pi l^2 & \\ \qquad + \frac{3}{4}l^2 - (\sqrt{3}\, l^2 + \frac{3\sqrt{3}}{2}a^2) \arccos \frac{\sqrt{3}\, a}{2l}, & \frac{\sqrt{3}}{2}a \leq l \leq a. \end{cases}$$

Right parallelepiped. If K is a right parallelepiped of length $a_1, a_2, \ldots, a_n$, then [16]

$$m(\ell, K) = \int_{S^{n-1}} \prod_{i=1}^{n} (a_i - \ell u \cdot e_i)_+ \, du,$$

where e_i are the unit coordinate vectors.

4. Chord power integrals and containment measures

The chord power integrals of a convex body K in $\mathbb{R}^n$ are defined by

$$I_\lambda(K) = \int_{K \cap L_1 \neq \phi} \sigma^\lambda \, dL_1,$$

where σ is the chord length $|K \cap L_1|$ of convex body K intercepted by line L_1. Chord power integrals and containment measures have the following relation,

$$I_\lambda(K) = \frac{1}{2}\lambda(\lambda-1)\int_0^\infty m(\ell, K)\ell^{\lambda-2}\,d\ell, \quad \lambda > 1.$$

The following are classical Crofton formulas,

$$I_1(K) = \frac{1}{2}V(K), \quad I_{n+1}(K) = \frac{n(n+1)}{2\alpha_n}V(K)^2.$$

There are inequalities between other chord power integrals and volume.

Theorem 4.1. ***(Blaschke, Wu and Ren)*** *If K is a convex body in $\mathbb{R}^n$, then*

$$\begin{aligned} I_\lambda(K) &\le I_\lambda(B) \quad 1 < \lambda < n+1, \\ I_\lambda(K) &\ge I_\lambda(B) \qquad\quad \lambda > n+1, \end{aligned}$$

where B is a ball that has the same volume as K. The equality of each inequality holds if and only if K is a ball.

The inequalities above were proved by Blaschke in $\mathbb{R}^2$, by Wu in $\mathbb{R}^3$, and by Ren in $\mathbb{R}^n$ (see [17,13,15]). Generalizations for flats were obtained by Schneider [19].

5. Sufficient conditions for containment of convex bodies and isoperimetric inequalities

If the areas and perimeters of two convex bodies in the plane are known, can one tell if one convex body can be contained in another by a rigid motion? Surprisingly, integral geometric methods give sufficient conditions to this containment problem. These sufficient conditions are closely related to the stability of isoperimetric inequalities. Necessary and sufficient conditions for the containment of two convex bodies by translation using circumscribed simplices were given by Lutwak [12].

Let K_0, K_1 be convex bodies in $\mathbb{R}^2$ with areas F_i and perimeters L_i. The fundamental kinematic formula of Blaschke is

$$\int_{\{g\in G_2:\, K_0\cap gK_1\neq\phi\}} dg = F_0 + F_1 + \frac{1}{2\pi}L_0L_1.$$

The Poincaré formula is

$$\int_{\{g\in G_2:\, \partial K_0\cap\partial(gK_1)\neq\phi\}} \sharp(\partial K_0 \cap \partial(gK_1))\,dg = \frac{2}{\pi}L_0L_1.$$

By these formulas, one can obtain a good lower bound of the containment measure,

$$\max\{m_1(K_0, K_1), m_1(K_1, K_0)\} \geq F_0 + F_1 - \frac{1}{2\pi} L_0 L_1.$$

If the containment measure $m_1(K_0, K_1) > 0$, then K_0 can be contained inside K_1 by a rigid motion. The inequality above gives a sufficient condition for the containment.

Theorem 5.1. ***(Hadwiger)*** *If K_0, K_1 are convex bodies in $\mathbb{R}^2$, then the following condition is sufficient for K_0 being contained in K_1 by a rigid motion*

$$4\pi F_0 - L_0 L_1 > (L_0^2 L_1^2 - 16\pi^2 F_0 F_1)^{\frac{1}{2}}.$$

The following sufficient condition for containment of convex bodies in $\mathbb{R}^2$ obtained by Delin Ren has more clear geometrical meaning.

Theorem 5.2. ***(Ren)*** *If K_0, K_1 are convex bodies in $\mathbb{R}^2$, then the following condition is sufficient for K_0 being contained in K_1 by a rigid motion,*

$$L_1 - L_0 > (\Delta_0 + \Delta_1)^{\frac{1}{2}},$$

where $\Delta_i = L_i^2 - 4\pi F_i$ are the isoperimetric deficits of K_i.

For convex bodies in $\mathbb{R}^n$, similar sufficient conditions are obtained by using methods of mixed volumes.

Theorem 5.3. ***(Zhang)*** *If K_0 and K_1 are convex bodies in $\mathbb{R}^n$ with volumes V_0 and V_1, then the following condition is sufficient for K_0 being contained in K_1 by a rigid motion,*

$$V_1^{\frac{1}{n}} - V_0^{\frac{1}{n}} > \left[\left(\frac{S_1 B_0}{2nV_0^{\frac{1}{n}}}\right)^{\frac{n}{n-1}} - V_1\right]^{\frac{1}{n}},$$

where B_0 is the mean width of K_0 and S_1 is the surface area of K_1.

When letting K_0 be the maximal ball inscribed in K_1 or letting K_1 be the minimal ball circumscribed of K_0, the inequality above yields Bonnesen-style isoperimetric inequalities.

Corollary 5.4. *If r and R are the inradius and outradius of a convex body K in $\mathbb{R}^n$ respectively, and B is the mean width of K, then*

$$\left(\frac{S}{n\omega_n}\right)^{\frac{n}{n-1}} - \frac{V}{\omega_n} \geq \left[\left(\frac{V}{\omega_n}\right)^{\frac{1}{n}} - r\right]^n,$$

$$\left(\frac{B}{2}\right)^{\frac{n}{n-1}} - \left(\frac{V}{\omega_n}\right)^{\frac{1}{n-1}} \geq \left(\frac{V}{\omega_n}\right)^{\frac{n}{n-1}} \left[\left(\frac{V}{\omega_n}\right)^{-\frac{1}{n}} - R^{-1}\right]^n,$$

where ω_n is the volume of the unit ball.

6. Sufficient conditions for containment of smooth convex bodies

For smooth convex bodies, the total curvatures of their boundaries can be used to derive lower bounds of containment measures, and thus to give sufficient conditions for containment of smooth convex bodies. The advantage of this integral-differential geometric approach is that it usually works for non-convex bodies. However, in this article, we deal with convex bodies only.

Let K be a convex body in $\mathbb{R}^n$ with C^2 smooth boundary ∂K, and let H be the mean curvature and dS be the surface area element of ∂K. Define the total mean curvature

$$M = \int_{\partial K} H \, dS,$$

and define the total square mean curvature

$$M^{(2)} = \int_{\partial K} H^2 \, dS.$$

The total mean curvature M and the quermassintegral W_2 has the relation,

$$M = nW_2.$$

6.1. *Sufficient conditions for containment of smooth convex bodies in $\mathbb{R}^3$*

Let K_0, K_1 be C^2 smooth convex bodies in $\mathbb{R}^3$. By the fundamental kinematic formula and Fenchel's inequality for the total curvature of space curves, it was shown in [24] that the containment measure of two smooth convex bodies in $\mathbb{R}^2$ has the following lower bound,

$$\max\{m_1(K_0, K_1), m_1(K_1, K_0)\} \geq V_0 + V_1 - \frac{1}{12\pi}(S_0 M_1 + S_1 M_0)$$
$$- \frac{1}{12\pi}\left(2S_0 S_1 [3(S_0 M_1^{(2)} + S_1 M_0^{(2)}) - 4\pi(S_0 + S_1) - 4M_0 M_1]\right)^{\frac{1}{2}}.$$

This inequality gives a sufficient condition for containment of two smooth convex bodies.

Theorem 6.1. ***(Zhang)*** *If K_0, K_1 are convex bodies with C^2 boundaries, then the following condition is sufficient for K_0 being contained in K_1 or K_1 being contained in K_0,*

$$12\pi(V_0+V_1)-(S_0M_1+S_1M_0) \\ -\left(2S_0S_1[3(S_0M_1^{(2)}+S_1M_0^{(2)})-4\pi(S_0+S_1)-4M_0M_1]\right)^{\frac{1}{2}}>0.$$

A different estimation for containment measure of smooth convex bodies in $\mathbb{R}^3$ was obtained by Zhou [29],

$$\max\{m_1(K_0,K_1),m_1(K_1,K_0)\}\geq V_0+V_1+\frac{1}{4\pi}(S_0M_1+S_1M_0) \\ -\frac{1}{8}\left(S_0S_1[3(S_0M_1^{(2)}+S_1M_0^{(2)})-4\pi(S_0+S_1)]\right)^{\frac{1}{2}}.$$

This inequality yields a sufficient condition for containment of two smooth convex bodies.

Theorem 6.2. ***(Zhou)*** *If K_0, K_1 are convex bodies with C^2 boundaries, then the following condition is sufficient for K_0 being contained in K_1 or K_1 being contained in K_0,*

$$8\pi(V_0+V_1)+2(S_0M_1+S_1M_0) \\ -\pi\left(S_0S_1[3(S_0M_1^{(2)}+S_1M_0^{(2)})-4\pi(S_0+S_1)]\right)^{\frac{1}{2}}>0.$$

Another sufficient condition for containment of two smooth convex bodies is the following

Theorem 6.3. ***(Zhou)*** *If K_0, K_1 are convex bodies with C^2 boundaries, then the following condition is sufficient for K_0 being contained in K_1 or K_1 being contained in K_0,*

$$8\pi(V_0+V_1)+2(S_0M_1+S_1M_0) \\ -\pi r\left(3(S_0M_1^{(2)}+S_1M_0^{(2)})-4\pi(S_0+S_1)\right)>0,$$

where r is the minimum of the radii of circumscribed balls of K_0 and K_1.

6.2. *Sufficient conditions for containment of smooth convex bodies in $\mathbb{R}^4$*

By using integral formulas of total integral of square mean curvature, Zhou [30] proved the following lower bound for containment measure of smooth convex bodies in $\mathbb{R}^4$,

$$\max\{m_1(K_0, K_1), m_1(K_1, K_0)\}$$
$$\geq V_0 + V_1 + \frac{7}{10}(B_0 S_1 + B_1 S_0) + \frac{3}{4\pi^2} M_0 M_1 - \frac{6}{5\pi^2}(M_0^{(2)} S_1 + M_1^{(2)} S_0).$$

Theorem 6.4. *(Zhou) If K_0, K_1 are convex bodies with C^2 boundaries, then the following condition is sufficient for K_0 being contained in K_1 or K_1 being contained in K_0,*

$$V_0+V_1+\frac{7}{10}(B_0S_1+B_1S_0)+\frac{3}{4\pi^2}M_0M_1-\frac{6}{5\pi^2}(M_0^{(2)}S_1+M_1^{(2)}S_0) > 0. \quad (6.1)$$

7. Geometric probabilities and containment measures

7.1. *Buffon's needle problem and its extensions*

Assume that the plane $\mathbb{R}^2$ is covered by parallel lines with distance d apart and a needle (line segment) of length $\ell < d$ is thrown at random in the plane. What is the probability that the needle hits a line? This well-known problem was posed and solved by Buffon in his Essai d'Arithmétique Morale (1777). Buffon's solutoin is that the probability equals $\frac{2\ell}{\pi d}$.

A lot of literature considered Buffon's needle problem. Ren's book has a detailed discussion about this problem. Buffon's needle problem was linked with containment measures by Ren. He considered extensions of Buffon's needle problem to lattices in $\mathbb{R}^n$.

Assume that $\mathbb{R}^n$ is divided by a lattice of regions $D_i = g_i D$, where D is a convex body and g_i are the elements of a discrete subgroup of G_n that keeps the lattice invariant. The union of the boundaries of D_i is called the boundary of the lattice.

Generalized Buffon needle problem (Ren). *Suppose that convex body D is the base region of a lattice in $\mathbb{R}^n$. A compact convex set K is thrown at random in $\mathbb{R}^n$, what is the probability that K hits the boundary of the lattice?*

When D is a parallelepiped with volume $V(D)$, Ren gave a solution to the generalized Buffon needle problem, which says that the probability is given by

$$p = 1 - \frac{m_1(K, D)}{V(D)}.$$

Ren's solution seems to be true for general lattices. But a proof is still to be given.

7.2. *Probe search*

Let K and D be convex bodies in $\mathbb{R}^n$, and N be a needle (line segment). Assume $K \subset D$. If a needle N is thrown inside of D, what is the probability that the needle hits K?

If the distance of K to the boundary of D is larger than the length of N, the probability is given by

$$p = \frac{m\{g \in G_n : K \cap gN \neq \phi\}}{m_1(N, D)},$$

where the numerator is the kinematic measure of N intersecting K that can be computed by the fundamental kinematic formula, and the denominator is the containment measure of N inside D that can be computed by using the chord projection function of D.

8. Isoperimetric problems of containment measures

By Theorems 3.3 and 3.5, it is easily seen that the containment measure $m(0, K)$ is the volume V of K, and the partial derivative $m'_\ell(0, K)$ is the surface area S of K by a constant factor,

$$m(0, K) = V, \qquad m'_\ell(0, K) = -\frac{\alpha_{n-1} S}{(n-1)\alpha_n}.$$

The containment measure $m(\ell, K)$ can be viewed as the amount of needles of length ℓ inside K. Thus, the following isoperimetric problem for containment measures is natural.

Isoperimetric problem for containment measures (Ren). *If the surface area S of K and the length ℓ, $0 < \ell < S/2$, of a line segment are given, what is the convex body K that attains the maximum of the containment measure $m(\ell, K)$?*

Instead of the surface area being fixed, if the diameter of convex body is fixed, the problem similar to the isoperimetric problem above is relatively easy to solve by using the Bieberbach inequality and Theorem 3.5, and has the following solution[25]:

If the diameter d of K and the length ℓ, $0 < \ell < d$, of a line segment are given, then the containment measure $m(\ell, K)$ attains its maximum if and only if K is a ball.

9. Sufficient conditions for containment of domains in space of constant curvature

For domains D_k $(k = i, j)$ in plane $\mathbf{X}^\kappa$ of constant curvature κ, i.e., the Euclidean plane $\mathbf{R}^2$ $(\kappa = 0)$, projective plane $\mathbf{PR}^2$ $(\kappa > 0)$ or hyperbolic plane $\mathbf{H}^2$ $(\kappa < 0)$, we introduce the symmetric isoperimetric deficit

$$\sigma(D_i, D_j) \equiv \Delta_i + \Delta_j - \kappa(F_i - F_j)^2, \tag{9.1}$$

where $\Delta_k = L_k^2 - 4\pi F_k + \kappa F_k^2$ is the isoperimetric deficit of D_k. Denote by G_2^κ the isometry of $\mathbf{X}^\kappa$. We have (see [7])

$$m\{g \in G_2^\kappa : gD_j \subset D_i \text{ or } gD_j \supset D_i\} \geq 2\pi(F_i + F_j) - L_i L_j - \kappa F_i F_j. \tag{9.2}$$

By letting $D_i \equiv D_j \equiv D$ leads to known generalized isoperimetric inequality

$$\Delta(D) = L^2 - 4\pi F + \kappa F^2 \geq 0, \tag{9.3}$$

with equality if and only if ∂D is a geodesic circle.

Proposition 9.1. ***(Grinberg, Ren and Zhou*** [7]***).*** *Let D_k $(k = i, j)$ be two domains with simple closed curves of perimeters L_k and areas F_k in the plane $\mathbf{X}^\kappa$. If*

$$2\pi(F_i + F_j) - L_i L_j - \kappa F_i F_j > 0 \tag{9.4}$$

then one of the domains contains another. If, in addition, $F_i \geq F_j$ or $L_i \geq L_j$, then D_i contains D_j.

Proposition 9.2. ***(Grinberg, Ren and Zhou*** [7]***).*** *Let D_k $(k = i, j)$ be two domains with simple closed curves of perimeters L_k and areas F_k in the plane $\mathbf{X}^\kappa$. Then either of the following conditions implies that D_i contains D_j.*

$$\sigma(D_i, D_j) < 0. \tag{9.5}$$

$$\sigma(D_i, D_j) \geq 0, \quad \text{and} \quad L_i - L_j > \sqrt{\sigma(D_i, D_j)}. \tag{9.6}$$

References

1. B. Y. Chen, Geometry of Submanifolds, Marcel Dekker. Inc., New York (1973).
2. B. Y. Chen, Total Mean Curvature and Submanifolds of Finite Type, World Scientific, Singapore (1984).
3. C. S. Chen, On the kinematic formula of square of mean curvature, Indiana Univ. Math. J., 22 (1972-73), 1163-1169.

4. S. S. Chern, On the kinematic formula in the euclidean space of n dimensions, Amer. J. Math., 74 (1952), 227-236.
5. S. S. Chern, On the kinematic formula in integral geometry, J. of Mathematics and Mechanics, 16 (1966), 101-118.
6. R. J. Gardner, Geometric Tomography, Cambridge University Press, Cambridge, 1995.
7. E. Grinberg, D. Ren and J. Zhou, The symmetric isoperimetric deficit and the containment problem in a plane of constant curvature, preprint.
8. D. Hug and R. Schneider, Kinematic and Crofton formulas of integral geometry: recent variants and extensions,
9. D. Klain and G-C. Rota, Introduction to Geometric Probability, Cambridge University Press, 1997
10. P. Li and S. T. Yau, A new conformal invariant and its applications to the Willmore conjecture and the first eigenvalue of compact surfaces, Invent. Math. 69 (1982), 269-291.
11. R. Li and G. Zhang, The kinematic measure of segment within some convex polygons and their applications to geometric probability problems, J. of Wuhan Iron and Steel University, 1 (1984), 106-128.
12. E. Lutwak, Containment and circumscribing simplices, Disc. Comput. Geom., 19(1998), 229-235.
13. D. Ren, Topics in Integral Geometry, World Scientific, Sigapore, 1994.
14. D. Ren, The generalized support function and its applications, Proceedings of the 1980 Beijing Symposium on differential geometry and differential equations, 1367-1378.
15. D. Ren, Two topics in integral geometry, Proceedings of the 1981 symposium on differential geometry and differential equations (Shanghai–Hefei), Sciences Press, Beijing, China, 1984, 309-333.
16. D. Ren and G. Zhang, Random convex sets in a lattice of parallelograms, Acta Mathematica Scientia, 11 (1991), 317-326.
17. L. A. Santaló, Integral Geometry and Geometric Probability. Addison-Wesley, Reading, Mass. (1976).
18. R. Schneider, Convex Bodies: The Brunn-Minkowski Theory, Cambridge Univ. Press, Cambridge (1993).
19. R. Schneider, Inequalities for random flats meeting a convex body, J. Appl. Prob., 22 (1985), 710-716.
20. R. Schneider and J.A. Wieacker, Integral geometry, Handbook of Convex Geometry (P.M. Gruber and J.M. Wills, Eds.), Vol. A, Elsevier, Amsterdam, 1993, 1349-1390.
21. W. Weil, Kinematic integral formulas for convex bodies, Contributions to Geometry (J. Tölke and J.M. Wills Eds.), Birkhäuser, Basel, 1979, 60-76.
22. T. Willmore, Total Curvature in Riemannian Geometry, Ellis Horwood, Chichester (1982).
23. L. Wu, Pairs of non-intersecting random flats meeting two convex bodies, Proceedings.
24. G. Zhang, A sufficient condition for one convex body containing another, Chin. Ann. of Math., 4 (1988), 447-451.

25. G. Zhang, Integral geometric inequalities, Acta Mathematica Sinica, 34 (1991), 72-90.
26. G. Zhang, Geometric inequalities and inclusion measures of convex bodies, Mathematika, 41 (1994), 95-116.
27. J. Zhou, kinematic formulas for mean curvature powers of hypersurfaces and Hadwiger's theorem in $\mathbf{R}^{2n}$, Trans. Amer. Math. Soc., 345 (1994), 243-262.
28. J. Zhou, On the Willmore deficit of convex surfaces, Lectures in Applied Mathematics of Amer. Math. Soc., 30 (1994), 279-287.
29. J. Zhou, When can one domain enclose another in $\mathbf{R}^3$? J. Austral. Math. Soc. (series A), 59 (1995), 266-272.
30. J. Zhou, The sufficient condition for a convex domain to contain another in $\mathbf{R}^4$, Proc. Amer. Math. Soc., 121 (1994), 907-913.
31. J. Zhou, A kinematic formula and analogous of Hadwiger's theorem in space, Contemporary Mathematics, 140 (1992), 159-167.
32. J. Zhou, Sufficient conditions for one domain to contain another in a space of constant curvature, Proc. Amer. Math. Soc., 126 (1998), 2797-2803.
33. J. Zhou, Kinematic formula for square mean curvature of hypersurfaces, Bull. of the Institute of Math., Academia Sinica, 22 (1994), 31-47.
34. J. Zhou, Notes on kinematic formula for square mean curvature, preprint.
35. J. Zhou, On the Willmore functional for hypersurface, preprint.

ON THE FLAG CURVATURE AND S-CURVATURE IN FINSLER GEOMETRY

XINYUE CHENG *

Department of Mathematics
Chongqing Institute of Technology
Chongqing 400050, P. R. China
E-mail: chengxy@cqit.edu.cn

Abstract. The flag curvature is an analogue of the sectional curvature in Riemannian geometry. In this paper, we study the relationship between the flag curvature and an important non-Riemannian quantity – S-curvature of Finsler metrics, and partially determine the flag curvature when Finsler metric is of scalar curvature and the S-curvature is isotropic . Furthermore, we classify the projectively flat Finsler metrics with isotropic S-curvature.

1. Notations and Definitions

A Finsler metric on a manifold M is a function $F : TM \to [0, \infty)$ which has the following properties:

(a) F is C^∞ on $TM\backslash\{0\}$;

(b) $F(x, \lambda y) = \lambda F(x, y), \quad \forall \lambda > 0$;

(c) For any tangent vector $y \in T_xM$, the following bilinear symmetric form $g_y : T_xM \times T_xM \to R$ is positive definite:

$$g_y(u, v) := \frac{1}{2} \frac{\partial^2}{\partial s \partial t} \left[F^2(x, y + su + tv)\right] |_{s=t=0}.$$

Let

$$g_{ij}(x, y) := \frac{1}{2} \left[F^2\right]_{y^i y^j}.$$

*supported by the national natural science foundation of china(10371138) and the science foundation of chongqing education committee.

By the homogeneity of F, we have the following

$$g_y(u,v) = g_{ij}(x,y)u^i v^j, \quad F(x,y) = \sqrt{g_{ij}(x,y)y^i y^j}.$$

Remark 1.1. We have the following special Finsler metrics

- Riemann metric: $F(x,y) = \sqrt{g_{ij}(x)y^i y^j}$, where g_{ij} are independent of $y \in T_xM$.
- Minkowski metric: $F(x,y) = \sqrt{g_{ij}(y)y^i y^j}$, where F is independent of $x \in M$.
- Randers metric (G. Randers, 1941): $F = \alpha + \beta$, where $\alpha = \sqrt{a_{ij}(x)y^i y^j}$ is a Riemannian metric and $\beta = b_i(x)y^i$ is a 1-form with $\|\beta\|_\alpha(x) := \sqrt{a^{ij}(x)b_i(x)b_j(x)}$ < 1 for any $x \in M$.

Now, we give a brief description of several geometric quantities in Finsler geometry. Let F be a Finsler metric on an n-dimensional manifold M. The geodesics of F are characterized by the equations:

$$\ddot{c}^i(t) + 2G^i\left(c(t), \dot{c}(t)\right) = 0,$$

where $G^i = G^i(x,y)$ are given by

$$G^i = \frac{1}{4}g^{il}\left\{[F^2]_{x^k y^l}y^k - [F^2]_{x^l}\right\},$$

which are called the *geodesic coefficients* of F.

The *Riemann curvature* $\mathbf{R}_y := R^i_k dx^k \bigotimes \frac{\partial}{\partial x^i}|_x : T_xM \to T_xM$ is a family of linear maps on tangent spaces, where

$$R^i_k = 2\frac{\partial G^i}{\partial x^k} - y^j\frac{\partial^2 G^i}{\partial x^j \partial y^k} + 2G^j\frac{\partial^2 G^i}{\partial y^j y^k} - \frac{\partial G^i}{\partial y^j}\frac{\partial G^j}{\partial y^k}. \tag{1}$$

For a flag $P = span\{y,u\} \subset T_xM$ with flagpole y, the *flag curvature* is defined by

$$\mathbf{K}(P,y) := \frac{g_y(u, \mathbf{R}_y(u))}{g_y(y,y)g_y(u,u) - g_y(y,u)^2}. \tag{2}$$

When F is Riemannian, $\mathbf{K}(P,y) = \mathbf{K}(P)$ is independent of flagpole $y \in P$, which is just the sectional curvature of P in Riemannian geometry. We say that F is *of scalar curvature* if for any $y \in T_xM$, the flag curvature $\mathbf{K}(P,y) = \mathbf{K}(x,y)$ is independent of P containing $y \in T_xM$. We say that F is *of constant curvature* if $\mathbf{K}(P,y)$=constant.

Let

$$\tau(x,y) := \ln\left[\frac{\sqrt{det(g_{ij}(x,y))}}{\sigma(x)}\right],$$

where

$$\sigma(x) := \frac{Vol(B^n)}{Vol\{(y^i) \in R^n | F(x,y) < 1\}}$$

and $\sigma(x)$ characterizes the Busemann-Hausdorff volume form of F. Then $\tau = \tau(x,y)$ is a scalar function on $TM\backslash\{0\}$, which is called the *distortion*[10].

The *mean Cartan torsion* $\mathbf{I} := I_i(x,y)dx^i : T_xM \to R$ is defined by

$$I_i(x,y) = \tau_{y^i}(x,y) = g^{jk}C_{ijk},$$

where $C_{ijk}(x,y) := \frac{1}{2}\frac{\partial g_{ij}}{\partial y^k}$ is *Cartan torsion.* According to Deicke's theorem(1953), F is Riemannian if and only if $\mathbf{I} = 0$, or equivalently, $\tau = \tau(x)$ at $x \in M$ [2].

Let

$$L_{ijk} = C_{ijk|m}y^m, \qquad J_i = I_{i|m}y^m,$$

where "|" denotes the horizontal covariant derivative with respect to the Chern connection. Then we obtain the *Landsberg curvature* $\mathbf{L} := L_{ijk}dx^i \otimes dx^j \otimes dx^k : T_xM \otimes T_xM \otimes T_xM \to R$ and the *mean Landsberg curvature* $\mathbf{J}_y = J_i(x,y)dx^i : T_xM \to R$.

Let

$$\mathbf{S} = \tau_{|m}y^m. \tag{3}$$

We call $\mathbf{S}$ the *S-curvature* [11]. We say S-curvature is *isotropic* if there is a scalar function $c(x)$ on M such that $\mathbf{S}(x,y) = (n+1)c(x)F(x,y)$. If $c(x) = constant$, we say that F has constant S-curvature. We have the following formula for S-curvature [11]

$$\mathbf{S} = \frac{\partial G^m}{\partial y^m} - y^m\frac{\partial(ln\sigma(x))}{\partial x^m}.$$

The S-curvature $\mathbf{S}$ measures the rate of change of the distortion τ along geodesics and the averages rate of change of (T_xM, F_x) in the direction $y \in T_xM$. Many known Finsler metrics of constant flag curvature actually have constant S-curvature (or vanishing S-curvature)[3 4 10].

It is easy to show that τ, $\mathbf{C}$, $\mathbf{I}$, $\mathbf{L}$, $\mathbf{J}$ and $\mathbf{S}$ all vanish for Riemannian metrics. Thus they are said to be non-Riemannian. The Riemannian quantities (such as Riemann curvature $\mathbf{R}$, flag curvature $\mathbf{K}$, Ricci curvature and Weyl curvature) describe the shape of a space, while non-Riemannian quantities describe the 'color' of the space.

A Finsler metric is said to be *locally projectively flat* if at any point there is a local coordinate system in which the geodesics are straight lines as point sets. The problem of characterizing and studying locally projectively flat Finsler metrics is known as *Hilbert's fourth problem.* Any locally projectively flat Finsler metric must be of scalar curvature. In particular, in Riemann geometry, we have known the following

Beltrami Theorem. A Riemann metric is locally projectively flat if and only if it is of constant curvature.

The local and global classification problem of locally projectively flat Finsler metrics with constant flag curvature has been solved at satisfactory level [12]. A Randers metric $F = \alpha + \beta$ is locally projectively flat if and only if α is a Riemann metric of constant sectional curvature and β is a closed 1-form ($d\beta = 0$).

2. Flag Curvature and S-Curvature

In this section, we would like to describe the effect of S-curvature for the geometric structure of Finsler metrics and introduce some recent developements in the study of the relationship between the flag curvature and S-curvature.

Firstly, we consider Finsler metrics of negative curvature. We should point out that there are already several global rigidity results on the metric structure of Finsler manifolds with $\mathbf{K} < 0$. For example, H. Akbar-Zadeh proves that every closed Finsler manifold of constant flag curvature with $\mathbf{K} < 0$ must be Riemannian[1]. Mo-Shen prove that every closed Finsler manifold of scalar curvature with $\mathbf{K} < 0$ must be of Randers type in dimension ≥ 3[8]. If we modify the condition that F is of constant (or scalar) curvature into the condition about S-curvature, we can see that S-curvature has great effect on the structure of Finsler metrics. In fact, we have the following

Theorem 2.1. [13] Let (M, F) be an n-dimensional closed Finsler manifold with constant S-curvature, i.e. $S = (n+1)cF$ for some constant c. If F has negative flag curvature, $K < 0$, then it must be Riemannian.

Secondly, we consider Finsler metrics of positive curvature. H. Akbar-Zadeh proves that every closed Finsler manifold (M, F) of constant flag curvature with $K > 0$ must be a topological sphere if M is simply connected[1].

Furthermore, Z. Shen proves that, if a reversible Finsler manifold (M, F) is complete simply connected and F is of constant curvature $K = 1$, then, for every $x \in M$, there is a unique point $x^* \in M$ with $d(x, x^*) = \pi$ and every normal geodesic issuing from x is closed with length 2π, passing through x^*[14]. Anyway, in general, it is very difficult to characterize the geometric structure of Finsler metric with positive constant curvature. However, we have the following

Theorem 2.2.[7] For a Finsler manifold (M, F), if (1) F is reversible, i.e. $F(x, -y) = F(x, y)$; (2) F is of positive constant curvature;(3) $\mathbf{S} = 0$, then F must be a Riemannian.

It is a difficult task to classify Finsler metrics of scalar curvature. All known Randers metrics of scalar curvature have isotropic S-curvature. Thus it is a natural idea to investigate Finsler metrics of scalar curvature with isotropic S-curvature. In 2003, X. Mo, Z. Shen and the author proved the following

Theorem 2.3.[6] Let (M, F) be an n-dimensional Finsler manifold of scalar curvature with flag curvature $\mathbf{K}(x, y)$. Suppose that

$$\mathbf{S} = (n+1)c(x)F(x, y). \tag{4}$$

Then there is a scalar function $\sigma(x)$ on M such that

$$\mathbf{K} = 3\frac{c_{x^m} y^m}{F(x, y)} + \sigma(x).$$

Further, $c = constant$ if and only if $\mathbf{K} = \mathbf{K}(x)$ is a scalar function on M.

Example 2.4. Given a Randers metric on $B^n(1/\sqrt{|a|})$ as follows

$$F = \frac{\sqrt{(|x|^2 < a, y > -2 < a, x >< x, y >)^2 + |y|^2(1 - |a|^2|x|^4)}}{1 - |a|^2|x|^4} - \frac{|x|^2 < a, y > -2 < a, x >< x, y >}{1 - |a|^2|x|^4},$$

where $a \in R^n$ is a constant vector and $y \in T_x B^n(1/\sqrt{|a|}) \cong R^n$. Then we can get the following

(i) $\mathbf{S} = (n+1) < a, x > F, \quad c(x) =< a, x >$;

(ii) $\mathbf{K} = 3\frac{<a,y>}{F} + 3 < a, x >^2 -2|a|^2|x|^2$, where

$$\sigma(x) = 3 < a, x >^2 -2|a|^2|x|^2.$$

(iii) F is not locally projrctively flat.

3. Funk Metric and Its Curvature Properties

Let $\phi = \phi(y)$ be a Minkowski norm on R^n and let

$$U_\phi := \{y \in R^n | \phi(y) < 1\}.$$

U_ϕ is called a *strongly convex domain.* For $0 \neq y \in T_x U_\phi \cong R^n$, define $\Theta(x, y) > 0$ by

$$x + \frac{y}{\Theta(x,y)} \in \partial U_\phi. \tag{5}$$

Equation (5) is equivalent to the following

$$\Theta(x, y) = \phi(y + \Theta(x, y)x). \tag{6}$$

A Finsler metric $\Theta = \Theta(x, y)$ defined in (5) or (6) is called the *Funk metric* on a strongly convex domain U_ϕ. Funk metric is the first known Finsler metric with constant flag curvature in Finsler geometry. In fact, Funk metric has many Riemannian (or non-Riemannian) curvature properties.

Lemma 3.1.[9] A Funk metric $\Theta = \Theta(x, y)$ on a strongly convex domain has the following properties:

(i) $\Theta_{x^k} = \Theta\Theta_{y^k}$.

(ii) Θ is locally projectively flat and $G^i = \frac{1}{2}\Theta y^i$.

Theorem 3.2. Let Θ denote Funk metric on a strongly convex domain in R^n, then

(1) Θ is a locally projectively flat Finsler metric;

(2) Θ is of constant S-curvature with $c = 1/2$:

$$\mathbf{S} = \frac{n+1}{2}\Theta;$$

(3) Θ has relatively isotropic Landsberg:

$$\mathbf{L} + \frac{1}{2}\Theta\mathbf{C} = 0;$$

(4) $\mathbf{K} = -\frac{1}{4}$ with $\sigma = -\frac{1}{4}$(cf. Theorem 2.3).

When $U_\phi = B^n$ is the standard unit ball in R^n, Funk metric

$$\Theta(x, y) = \frac{\sqrt{|y|^2 - (|x|^2|y|^2 - <x, y>^2)}}{1 - |x|^2} + \frac{<x, y>}{1 - |x|^2},$$

It is just a Randers metric on B^n.

In 2003, Z. Shen and the author classified projectively flat Randers metrics with isotropic S-curvature [6]. It is a natural problem to study and characterize projectively flat Finsler metrics with isotropic S-curvature.

We want to know whether or not there are other types of projectively flat Finsler metrics of isotropic S-curvature.

Theorem 3.3. [5] Let $F = F(x, y)$ be a locally projectively flat Finsler metric on a simply connected open subset $U \subset R^n$. Suoopse that F has isotropic S-curvature, i.e.

$$\mathbf{S} = (n+1)c(x)F, \tag{7}$$

where $c = c(x)$ is a scalar function on M. Then

$$\mathbf{K} = 3\frac{c_{x^m} y^m}{F(x,y)} + \sigma(x) \tag{8}$$

and F is determined as follows.

(a) If $\mathbf{K} \neq -c^2 + \frac{c_{x^m} y^m}{F}$ at every point $x \in U$, then $F = \alpha + \beta$ is a projectively flat Randers metric with isotropic S-curvature on U;

(b) If $\mathbf{K} \equiv -c^2 + \frac{c_{x^m} y^m}{F}$ on U, then c is a constant, and either F is locally Minkowskian ($c = 0$) or up to a scaling, F is locally isometric to the metric

$$\Theta_a = \Theta(x, y) + \frac{< a, y >}{1 + < a, x >} \qquad (c = \frac{1}{2})$$

$$or \quad \bar{\Theta}_a = \Theta_a(x, -y) \qquad (c = -\frac{1}{2}),$$

where $a \in R^n$ is a constant vector and $\Theta = \Theta(x, y)$ is a Funk metric defined by

$$\Theta_{x^k}(x, y) = \Theta(x, y)\Theta_{y^k}(x, y).$$

Acknowledgements. The author would like to thank Professor Z. Shen, Professor J. Zhou and Professor G. Zhang for their great help and hospitality. The author also would like to thank the referees for their valuable suggestion.

References

1. H. Akbar-Zadeh, Sur les espaces de Finsler *á* courbures sectionnelles constantes, Bull.Acad. Roy. Bel. CI, Sci, 5e Série-Tome LXXXIV(1988), 281-322.
2. D. Bao, S. S. Chern and Z. Shen, An Introduction to Riemann-Finsler Geometry, Springer, 2000.
3. D. Bao and C. Robles, On Randers metrics of constant curvature, Rep. on Math. Phys.**51**(2003), 9-42.

4. X. Chen and Z. Shen, Randers metrics with special curvature properties, Osaka J. of Math.**40**(2003), 87-101.
5. X. Cheng and Z. Shen, Projectively flat Finsler metrics with almost isotropic S-curvature, Acta Mathematica Scientia(to appear).
6. X. Chen, X. Mo and Z. Shen, On the flag curvature of Finsler metrics of scalar curvature,J. of London Math. Soc., (2)**68**(2003), 762-780.
7. C.-W Kim and J.-W Yim, Finsler manifolds with positive constant flag curvature, Geo. Dedicata, **98**(2003), 47-56.
8. X. MO and Z. Shen, On negatively curved Finsler manifolds of scalar curvature, Canadian Math.Bull., (to appear).
9. T. Okada, On models of projectively flat Finsler spaces of constant negative curvature,Tensor, N. S.**40**(1983), 117-123.
10. Z. Shen, Lectures on Finsler Geometry, World Scientific, Singapore, 2001.
11. Z. Shen, Differential Geometry of Spray and Finsler Spaces, Kluwer Academic Publishers, 2001.
12. Z. Shen, Projectively flat Finsler metrics of constant flag curvature, Trans. of Amer. Math. Soc., **355**(4)(2002), 1713-1728.
13. Z. Shen, Nonpositively curved Finsler manifolds with constant S-curvature,preprint, 2003.
14. Z. Shen, Finsler spaces of constant positive curvature, Contemporary Math.**196**(1996), 83-92.

DOUBLE CHORD-POWER INTEGRALS OF A CONVEX BODY AND THEIR APPLICATIONS

PENG XIE

Department of Mathematics Huazhong University of Science and Technology, Wuhan, Hubei 430074, China
E-mail: xiepeng@mail.hust.edu.cn

JUN JIANG

Department of Mathematics Wuhan University of Science and Technology, Wuhan, Hubei 430081, China
E-mail: jiangjunwkd@eyou.com

(*Dedicated to Professor Delin Ren on the occasion of his 70th birthday.*)

In this paper, the concept of double chord-power integrals of a convex body is introduced. The properties of double chord-power integrals and applications in geometric probability are given.

1. Introduction

Let K be a convex body in $\mathbb{R}^2$, and σ the chord intersected by a random line G with K. Consider integrals

$$I_n(K) = \int_{G \bigcap K \neq \emptyset} \sigma^n dG$$

where n is a nonnegative integer and dG is the invariant density of lines [3,5]. The integral $I_n(K)$ is called the nth chord-power integral of K [3].

Let L be the perimeter and F the area of the convex body K. There are integral formulas [3],

$$I_0(K) = L, \tag{1}$$

$$I_1(K) = \pi F, \tag{2}$$

and Crofton's formula [5],

$$I_3(K) = 3F^2. \tag{3}$$

There are also geometric inequalities of the chord-power integrals [3],

$$I_2(K) \leq \frac{2^4}{3\pi^2} I_1^{\frac{3}{2}}(K) \tag{4}$$

$$I_n(K) \geq \frac{2 \cdot 4 \cdots n}{3 \cdot 5 \cdots (n+1)} \pi (\frac{2}{\pi} I_1^{\frac{1}{2}}(K))^{n+1}, \quad n = 4, 6, 8, \cdots \tag{5}$$

$$I_n(K) \geq \frac{1 \cdot 3 \cdots n}{2 \cdot 4 \cdots (n+1)} \frac{\pi^2}{2} (\frac{2}{\pi} I_1^{\frac{1}{2}}(K))^{n+1}, \quad n = 3, 5, 7, \cdots . \tag{6}$$

By (1) and (2), the classical isoperimetric inequality $L^2 \geq 4\pi F$ can be written as

$$I_0^2(K) \geq 4 I_1(K).$$

Therefore, it is interesting and meaningful to investigate the relations among those I_n and inequalities of I_n. The inequalities (4)-(6) were proved by T. J. Wu, and their generalizations in higher dimensions were obtained by Delin Ren [3,4,7]. Dual kinematic formulas of the chord power integrals were given by Zhang [8].

Let K be a convex body in $\mathbb{R}^2$ and $z \in K$. The radial function ρ_K of K with respect to z is defined by $\rho_K(z,u) = \max\{c > 0 : z + cu \in K\}$ for $u \in S^1$. There is the formula for the dual quermassintegrals [2],

$$\tilde{W}_{2-r}(K, z) = \frac{1}{2} \int_{S^1} \rho_K^r(z,u) du, \quad r \in \mathbb{R}$$

The chord-power integrals of a convex body are related to dual quermass-integrals. There is the formula [8],

$$I_n(K) = n \int_{z \in K} \tilde{W}_{3-n}(K, z) dz.$$

In this paper, we present an extension of the chord-power integrals. We introduce the concept of *double chord-power integrals* of a convex body. The chord power integrals are special cases of our double chord-power integrals. We show properties of the double chord-power integrals and calculate special cases of the double chord-power integrals when the convex body is a circle. Applications of the double chord-power integrals in geometric probability are given.

2. Double chord-power integrals of a convex body

Definition. *Let K be a convex body in $\mathbb{R}^2$, and σ_1, σ_2 be respectively the chords intersected by random lines G_1, G_2. The double chord-power integrals are defined as*

$$I_{mn}(K) = \int_{G_1 \bigcap G_2 \in K} \sigma_1^m \sigma_2^n \, dG_1 dG_2$$

where m, n are nonnegative integers.

Proposition 1. *Let K be a convex body in $\mathbb{R}^2$. Then we have*

$$I_{mn}(K) = \frac{1}{4} \int_{z \in K} \tilde{W}_{mn}(K, z) dz, \tag{7}$$

where

$$\tilde{W}_{mn}(K, z) = \int_{(u_1, u_2) \in S^1 \times S^1} w_{mn}(z, u_1, u_2) du_1 du_2;$$

$$w_{mn}(z, u_1, u_2) = |u_1 \wedge u_2| (\rho_K(z, u_1) + \rho_K(z, -u_1))^m (\rho_K(z, u_2) + \rho_K(z, -u_2))^n,$$

$|u_1 \wedge u_2| = |\sin(\alpha_1 - \alpha_2)|$, *$\alpha_i$ are the angles of u_i with the x-axis (i=1, 2).*

Proof. Let G_i^* be the oriented lines whose directions are $u_i \in S^1, i = 1, 2$. If z is the intersection of G_1 and G_2, there is the density formula for pairs of lines [3]

$$dG_1^* dG_2^* = |\sin(\alpha_1 - \alpha_2)| dz d\alpha_1 d\alpha_2 = |u_1 \wedge u_2| dz du_1 du_2.$$

Therefore, by

$$\sigma_i = \rho_K(z, u_i) + \rho_K(z, -u_i),$$

we obtain

$$\begin{aligned} I_{mn}(K) &= \int_{G_1 \bigcap G_2 \in K} \sigma_1^m \sigma_2^n dG_1 dG_2 = \frac{1}{4} \int_{G_1^* \bigcap G_2^* \in K} {\sigma_1}^m {\sigma_2}^n dG_1^* dG_2^* \\ &= \frac{1}{4} \int_{z \in K} \int_{(u_1, u_2) \in S^1 \times S^1} |u_1 \wedge u_2| \sigma_1^m \sigma_2^n dz du_1 du_2 \\ &= \frac{1}{4} \int_{z \in K} \tilde{W}_{mn}(K, z) dz. \end{aligned}$$

Proposition 2. *Double chord-power integrals are symmetric,*

$$I_{mn}(K) = I_{nm}(K). \tag{8}$$

Proof.

$$\begin{aligned} I_{mn}(K) &= \int_{G_1 \bigcap G_2 \in K} \sigma_1^m \sigma_2^n dG_1 dG_2 \\ &= \int_{G_1 \bigcap K \neq \emptyset} \sigma_1^m dG_1 \int_{G_2 \bigcap (G_1 \bigcap K) \neq \emptyset} \sigma_2^n dG_2 \\ &= \int_{G_2 \bigcap K \neq \emptyset} \sigma_2^m dG_2 \int_{G_1 \bigcap (G_2 \bigcap K) \neq \emptyset} \sigma_1^n dG_1 = I_{nm}(K). \end{aligned}$$

The following proposition shows that chord-power integrals are special cases of double chord-power integrals.

Proposition 3. *The chord-power integrals and the double chord-power integrals have the relations,*

$$I_{0n}(K) = I_{n0}(K) = 2I_{n+1}(K). \tag{9}$$

In particular,

$$I_{00}(K) = 2I_1(K) = 2\pi F, \tag{10}$$

$$I_{01}(K) = I_{10}(K) = 2I_2(K). \tag{11}$$

Proof. By Proposition 2, there is $I_{0n}(K) = I_{n0}(K)$. By (1), we have

$$\begin{aligned} I_{n0}(K) &= \int_{G_1 \bigcap G_2 \in K} \sigma_1^n dG_1 dG_2 = \int_{G_1 \bigcap K \neq \emptyset} \sigma_1^n dG_1 \int_{G_2 \bigcap (G_1 \bigcap K) \neq \emptyset} dG_2 \\ &= \int_{G_1 \bigcap K \neq \emptyset} 2\sigma_1^{n+1} dG_1 = 2I_{n+1}(K). \end{aligned}$$

Proposition 4. *The double chord-power integrals have the inequalities,*

$$I_{mn}(K) \leq I_{2m+1}(K) + I_{2n+1}(K). \tag{12}$$

In particular,

$$I_{mm}(K) \leq 2I_{2m+1}(K), \tag{13}$$

$$I_{11}(K) \leq 2I_3(K) = 6F^2. \tag{14}$$

Proof. The inequality follows from

$$\begin{aligned}
0\leq&\int_{G_1\bigcap G_2\in K}(\sigma_1^m-\sigma_2^n)^2dG_1dG_2\\
=&\int_{G_1\bigcap G_2\in K}(\sigma_1^{2m}-2\sigma_1^m\sigma_2^n+\sigma_2^{2n})dG_1dG_2\\
=&\int_{G_1\bigcap G_2\in K}\sigma_1^{2m}dG_1dG_2-2\int_{G_1\bigcap G_2\in K}\sigma_1^m\sigma_2^n dG_1dG_2\\
&+\int_{G_1\bigcap G_2\in K}\sigma_2^{2n}dG_1dG_2\\
=&2I_{2m+1}(K)-2I_{mn}(K)+2I_{2n+1}(K).
\end{aligned}$$

Proposition 5. *Let B be the circle of radius R centered at the origin, then*

$$I_{mn}(B)=2^{m+n+2}\pi\int_0^R rJdr, \tag{15}$$

where

$$J=\int_0^\pi (R^2-r^2\sin^2\varphi_1)^{\frac{m}{2}}d\varphi_1\int_0^{\varphi_1}(R^2-r^2\sin^2\varphi_2)^{\frac{n}{2}}\sin(\varphi_1-\varphi_2)d\varphi_2.$$

Proof. Let z be the intersection of the lines G_1, G_2, $|z|=r$. For $i=1,2$, let φ_i be the angles between z and u_i, α_i the angles between u_i and the x-axis, and θ the angle of z with the x-axis. By

$$\begin{aligned}
\rho_B(z,u_i)&=-zu_i+\sqrt{R^2-(z\wedge u_i)^2},\\
\alpha_i&=\frac{\pi}{2}+\theta-\varphi_i,
\end{aligned}$$

we have

$$\sigma_i=\rho_B(z,u_i)+\rho_B(z,-u_i)=2\sqrt{R^2-(z\wedge u_i)^2}=2\sqrt{R^2-r^2\sin^2\varphi_i}.$$

Therefore

$$\begin{aligned}
I_{mn}(B)=&\int_{G_1\bigcap G_2\in B}\sigma_1^m\sigma_2^n dG_1dG_2=\frac{1}{4}\int_{G_1\bigcap G_2\in B}\sigma_1^m\sigma_2^n dG_1^*dG_2^*\\
=&2^{m+n-2}\int_0^{2\pi}d\theta\int_0^R rdr\int_0^{2\pi}(R^2-r^2\sin^2\varphi_1)^{\frac{m}{2}}d\alpha_1\\
&\int_0^{2\pi}(R^2-r^2\sin^2\varphi_2)^{\frac{n}{2}}|\sin(\alpha_1-\alpha_2)|d\alpha_2\\
=&2^{m+n+1}\pi\int_0^R r(J+J^*)dr,
\end{aligned}$$

where

$$J^* = \int_0^{\pi} (R^2 - r^2 \sin^2 \varphi_1)^{\frac{m}{2}} d\varphi_1 \int_{\varphi_1}^{\pi} (R^2 - r^2 \sin^2 \varphi_2)^{\frac{n}{2}} \sin(\varphi_2 - \varphi_1) d\varphi_2.$$

Let $\beta = \pi - \varphi_2$, $\gamma = \pi - \varphi_1$. We have

$$\begin{aligned} J^* =& \int_0^{\pi} (R^2 - r^2 \sin^2 \varphi_1)^{\frac{m}{2}} d\varphi_1 \int_0^{\pi-\varphi_1} (R^2 - r^2 \sin^2 \beta)^{\frac{n}{2}} \sin(\pi - \beta - \varphi_1) d\beta \\ =& \int_0^{\pi} (R^2 - r^2 \sin^2 \gamma)^{\frac{m}{2}} d\gamma \int_0^{\gamma} (R^2 - r^2 \sin^2 \beta)^{\frac{n}{2}} \sin(\gamma - \beta) d\beta \\ =& J. \end{aligned}$$

Therefore,

$$I_{mn}(B) = 2^{m+n+1}\pi \int_0^R r(J + J^*)dr = 2^{m+n+2}\pi \int_0^R rJdr.$$

Proposition 6. *Let B be the circle of radius R centered at the origin, then*

$$I_{mm}(B) = 2^{2m+4}\pi \int_0^R r\widehat{J}dr, \tag{16}$$

where

$$\widehat{J} = \int_0^{\frac{\pi}{2}} (R^2 - r^2 \sin^2 \varphi_1)^{\frac{m}{2}} \sin\varphi_1 d\varphi_1 \int_0^{\varphi_1} (R^2 - r^2 \sin^2 \varphi_2)^{\frac{m}{2}} \cos\varphi_2 d\varphi_2.$$

Proof. Let $J = J_1 + J_2$, where

$$\begin{aligned} J_1 =& \int_0^{\pi} (R^2 - r^2 \sin^2 \varphi_1)^{\frac{m}{2}} \sin\varphi_1 d\varphi_1 \int_0^{\varphi_1} (R^2 - r^2 \sin^2 \varphi_2)^{\frac{m}{2}} \cos\varphi_2 d\varphi_2 \\ =& 2\int_0^{\frac{\pi}{2}} (R^2 - r^2 \sin^2 \varphi_1)^{\frac{m}{2}} \sin\varphi_1 d\varphi_1 \int_0^{\varphi_1} (R^2 - r^2 \sin^2 \varphi_2)^{\frac{m}{2}} \cos\varphi_2 d\varphi_2 \\ =& 2\widehat{J} \end{aligned}$$

and for $\gamma = \pi - \varphi_1, \beta = \pi - \varphi_2$,

$$\begin{aligned} J_2 =& -\int_0^{\pi} (R^2 - r^2 \sin^2 \varphi_1)^{\frac{m}{2}} \cos\varphi_1 d\varphi_1 \int_0^{\varphi_1} (R^2 - r^2 \sin^2 \varphi_2)^{\frac{m}{2}} \sin\varphi_2 d\varphi_2 \\ =& -\int_0^{\pi} (R^2 - r^2 \sin^2 \varphi_2)^{\frac{m}{2}} \sin\varphi_2 d\varphi_2 \int_{\varphi_2}^{\pi} (R^2 - r^2 \sin^2 \varphi_1)^{\frac{m}{2}} \cos\varphi_1 d\varphi_1 \\ =& \int_0^{\pi} (R^2 - r^2 \sin^2 \beta)^{\frac{m}{2}} \sin\beta d\beta \int_0^{\beta} (R^2 - r^2 \sin^2 \gamma)^{\frac{m}{2}} \cos\gamma d\gamma = 2\widehat{J}. \end{aligned}$$

Therefore by (15), we have

$$\begin{aligned}I_{mm}(B) =& 2^{2m+2}\pi \int_0^R rJdr \\ =& 2^{2m+2}\pi \int_0^R r(J_1+J_2)dr = 2^{2m+4}\pi \int_0^R r\widehat{J}dr.\end{aligned}$$

Proposition 7. *Let B be the circle of radius R centered at the origin, then*

$$I_{22}(B) = 41(\frac{2}{3})^2 \pi^2 R^6. \tag{17}$$

Proof. By (16), we have

$$\begin{aligned}I_{22}(B) =& 2^8\pi \int_0^R rdr \int_0^{\frac{\pi}{2}} (R^2 - r^2\sin^2\varphi_1)\sin\varphi_1 d\varphi_1 \\ & \int_0^{\varphi_1} (R^2 - r^2\sin^2\varphi_2)\cos\varphi_2 d\varphi_2 \\ =& 2^8\pi \int_0^R rdr \int_0^{\frac{\pi}{2}} (R^4\sin^2\varphi_1 - \frac{4}{3}R^2r^2\sin^4\varphi_1 + \frac{1}{3}r^4\sin^6\varphi_1)d\varphi_1 \\ =& 41(\frac{2}{3})^2\pi^2R^6.\end{aligned}$$

Proposition 8. *Let B be the circle of radius R centered at the origin, then*

$$I_{11}(B) = (\frac{5}{2} + \frac{1}{3}\pi^2)\pi^2R^4. \tag{18}$$

Proof. Since

$$\begin{aligned}\widehat{J} =& \int_0^{\frac{\pi}{2}} \sin\varphi_1\sqrt{R^2 - r^2\sin^2\varphi_1}d\varphi_1 \int_0^{\varphi_1} \cos\varphi_2\sqrt{R^2 - r^2\sin^2\varphi_2}d\varphi_2 \\ =& \frac{1}{2}\int_0^{\frac{\pi}{2}} \sin^2\varphi_1(R^2 - r^2\sin^2\varphi_1)d\varphi_1 \\ &+ \frac{R^2}{2r}\int_0^{\frac{\pi}{2}} \sin\varphi_1\sqrt{R^2 - r^2\sin^2\varphi_1}\arcsin\frac{r\sin\varphi_1}{R}d\varphi_1 \\ =& \frac{\pi}{8}(R^2 - \frac{3}{4}r^2) + \frac{R^2}{2r^2}\int_0^r \frac{t\sqrt{R^2-t^2}}{\sqrt{r^2-t^2}}\arcsin\frac{t}{R}dt,\end{aligned}$$

by (16), we have

$$
\begin{aligned}
I_{11}(B) &= 2^6\pi \int_0^R r\widehat{J}dr \\
&= 2^6\pi \int_0^R \frac{\pi}{8}(R^2 r - \frac{3}{4}r^3)dr + 2^5\pi R^2 \int_0^R \frac{1}{r}dr \int_0^r \frac{t\sqrt{R^2-t^2}}{\sqrt{r^2-t^2}} \arcsin\frac{t}{R}dt \\
&= \frac{5}{2}\pi^2 R^4 + 2^5\pi R^2 \int_0^R t\sqrt{R^2-t^2}\arcsin\frac{t}{R}dt \int_t^R \frac{1}{r\sqrt{r^2-t^2}}dr \\
&= \frac{5}{2}\pi^2 R^4 + 2^5\pi R^2 \int_0^R \sqrt{R^2-t^2}\arcsin\frac{t}{R}(\frac{\pi}{2} - \arcsin\frac{t}{R})dt \\
&= \frac{5}{2}\pi^2 R^4 + 2^5\pi R^4 \int_0^{\frac{\pi}{2}} (\frac{\pi}{2}x - x^2)\cos^2 x dx = (\frac{5}{2} + \frac{1}{3}\pi^2)\pi^2 R^4.
\end{aligned}
$$

3. Applications to a problem of geometric probability

A problem of geometric probability: *If K is a convex body in $\mathbb{R}^2$ and N_1, N_2 are random line segments intersecting with K that are contained in lines G_1, G_2, what is the probability that the lines G_1, G_2 intersect at a point in K?*

We will give a solution of this problem by using the double chord-power integrals.

Let L be the perimeter and F the area of the convex body K . Let N_i be the random line segment of length l_i (i=1,2), and let G_i be the line containing N_i (i=1,2). The invariant density of N_i can be chosen as $dN_i = dG_i dt_i$, where dG_i is the invariant density of random lines and t_i is the coordinate of N_i on the line G_i [3]. We have the following density formula,

$$dN_1 dN_2 = dG_1 dG_2 dt_1 dt_2.$$

Lemma 1. *If K is a convex body in $\mathbb{R}^2$ and N_1, N_2 are random line segments intersecting with K that are contained in lines G_1, G_2, then*

$$\int_{N_1 \bigcap K \neq \emptyset, N_2 \bigcap K \neq \emptyset, G_1 \bigcap G_2 \in K} dN_1 dN_2 = 2\pi F l_1 l_2 + 2(l_1+l_2) I_2(K) + I_{11}(K).$$

Proof. By (10) and (11), we have

$$\begin{aligned}
&\int_{N_1 \bigcap K\neq\emptyset, N_2 \bigcap K\neq\emptyset, G_1 \bigcap G_2 \in K} dN_1 dN_2 \\
=&\int_{G_1 \bigcap G_2 \in K} dG_1 dG_2 \int_0^{l_1+\sigma_1} dt_1 \int_0^{l_2+\sigma_2} dt_2 \\
=&\int_{G_1 \bigcap G_2 \in K} (l_1+\sigma_1)(l_2+\sigma_2) dG_1 dG_2 \\
=&l_1 l_2 \int_{G_1 \bigcap G_2 \in K} dG_1 dG_2 + l_1 \int_{G_1 \bigcap G_2 \in K} \sigma_2 dG_1 dG_2 \\
&+ l_2 \int_{G_1 \bigcap G_2 \in K} \sigma_1 dG_1 dG_2 + \int_{G_1 \bigcap G_2 \in K} \sigma_1 \sigma_2 dG_1 dG_2 \\
=&l_1 l_2 I_{00}(K) + (l_1+l_2) I_{01}(K) + I_{11}(K) \\
=&2\pi F l_1 l_2 + 2(l_1+l_2) I_2(K) + I_{11}(K).
\end{aligned}$$

Lemma 2. *If K is a convex body in $\mathbb{R}^2$ and N_1, N_2 are random line segments intersecting with K, then*

$$\int_{N_1 \bigcap K\neq\emptyset, N_2 \bigcap K\neq\emptyset} dN_1 dN_2 = (\pi F + l_1 L)(\pi F + l_2 L).$$

Proof. By (1) and (2), we have

$$\begin{aligned}
&\int_{N_1 \bigcap K\neq\emptyset, N_2 \bigcap K\neq\emptyset} dN_1 dN_2 \\
=&\int_{G_1 \bigcap K\neq\emptyset} dG_1 \int_{G_2 \bigcap K\neq\emptyset} dG_2 \int_0^{l_1+\sigma_1} dt_1 \int_0^{l_2+\sigma_2} dt_2 \\
=&\int_{G_1 \bigcap K\neq\emptyset} (\sigma_1+l_1) dG_1 \int_{G_2 \bigcap K\neq\emptyset} (\sigma_2+l_2) dG_2 \\
=&(I_1(K) + l_1 I_0(K))(I_1(K) + l_2 I_0(K)) \\
=&(\pi F + l_1 L)(\pi F + l_2 L).
\end{aligned}$$

Theorem. *Let K be a convex body in $\mathbb{R}^2$. Let N_1, N_2 be random line segments intersecting with K that are respectively contained in lines G_1, G_2. Then the probability that the lines G_1, G_2 intersect at a point in K is given by*

$$p(K) = \frac{2\pi F l_1 l_2 + 2(l_1 + l_2) I_2(K) + I_{11}(K)}{(\pi F + l_1 L)(\pi F + l_2 L)}. \tag{19}$$

Corollary 1. *If $l_1 = l_2 = l$, then*

$$p(K) = \frac{2\pi F l^2 + 4 l I_2(K) + I_{11}(K)}{(\pi F + lL)^2}. \tag{20}$$

In particular, for a ball, by (18) , *we have*

$$p(B) = \frac{2\pi^2 l^2 R^2 + \frac{2^6}{3}\pi l R^3 + (\frac{5}{2} + \frac{\pi^2}{3})\pi^2 R^4}{(\pi^2 R^2 + 2\pi l R)^2}. \tag{21}$$

Corollary 2. *If $l_1 = l_2 = 0$, then*

$$p(K) = \frac{I_{11}(K)}{(\pi F)^2} \leq \frac{6}{\pi^2}. \tag{22}$$

In fact, the case $l_1 = l_2 = 0$ gives a solution to the following problem: *Pick two points randomly in a convex body K and draw two lines passing through the two points, what is the probability that the intersection of the two lines belongs to K?*

Let us give a direct calculation for this problem. Let G_i^* are oriented lines whose direction is $u_i \in S$. Let $z_i \in K$ and G_i^* passes through z_i, Denote by t_i the distance of z_i to the boundary of K along the direction u_i, α_i the angles between u_i and the x-axis. Then we have [3,8]

$$dG_i^* dt_i = dz_i du_i, i = 1, 2$$

$$\begin{aligned} p(K) &= \frac{\int_{z_i\in K, u_i\in S, G_1\bigcap G_2\in K} dz_1dz_2du_1du_2}{\int_{z_i\in K,u_i\in S} dz_1dz_2du_1du_2} \\ &= \frac{\int_{G_1^*\bigcap G_2^*\in K} dG_1^*dG_2^* \int_0^{\sigma_1} dt_1 \int_0^{\sigma_2} dt_2}{\int_{z_1\in K} dz_1 \int_{z_2\in K} dz_2 \int_0^{2\pi} d\alpha_1 \int_0^{2\pi} d\alpha_2} \\ &= \frac{4\int_{G_1\bigcap G_2\in K} \sigma_1\sigma_2 dG_1dG_2}{4\pi^2 \int_{z_1\in K} dz_1 \int_{z_2\in K} dz_2} \\ &= \frac{I_{11}(K)}{(\pi F)^2}. \end{aligned}$$

In particular, for a ball, by (18), we have

$$p(B) = \frac{1}{3} + \frac{5}{2\pi^2}. \tag{23}$$

Corollary 3. *If $l_1 = l_2 = l$, then*

$$\lim_{l\to\infty} p(K) = \frac{2\pi F}{L^2} \leq \frac{1}{2} \tag{24}$$

with equality if and only if K is a ball [6].

Proof. If $l_1 = l_2 = l$, then

$$\lim_{l\to\infty} p(K) = \lim_{l\to\infty} \frac{2\pi F l^2 + 4lI_2(K) + I_{11}(K)}{(\pi F + lL)^2} = \frac{2\pi F}{L^2}.$$

In fact, the limit case $l_1 = l_2 = l \to \infty$ gives a solution to the following problem: *If G_1 and G_2 are random lines intersect with a convex body K, what is the probability that the intersection of G_1 and G_2 belongs to the convex body K?* A direct calculation gives,

$$p = \frac{\int_{G_1\bigcap G_2\in K} dG_1dG_2}{\int_{G_1\bigcap K\neq\emptyset} dG_1 \int_{G_2\bigcap K\neq\emptyset} dG_2} = \frac{I_{00}(K)}{I_0(K)I_0(K)} = \frac{2\pi F}{L^2}.$$

4. Problems

Chord-power integrals are important geometric invariants. They are useful in geometric tomography [1]. The double chord-power integrals may also be useful in geometric tomography. It would be interesting to obtain the properties of chord-power integrals to double chord-power integrals. We just state a few problems here.

Problem 1. Are there the inequalities among double chord-power integrals similar to inequalities among chord-power integrals (4)-(6)?

Problem 2. Is a convex body in the plane uniquely determined by its double chord-power integrals?

Problem 3. Is there any kinematic formula associated with the double chord-power integrals?

Acknowledgments

We would like to thank Professors Delin Ren, Jiazu Zhou, and Gaoyong Zhang, who taught us integral geometry. We appreciate very much for their help and encouragement for years.

References

1. R. Gardner, Geometric Tomography, *Cambridge University Press.* (1995).
2. E. Lutwak, Dual mixed volumes, *Pacific J. Math.* **58**, 531-538 (1975).
3. D. L. Ren, Topics in Integral Geometry, *World Scientific, Singapore.* (1994).
4. D. L. Ren, Two topics integral geometry, *Proceedings of the 1981 symposium on differential geometry and differential equations (ShangHai-HeFei), Science Press, Beijing, China.* 309-333 (1984).
5. L. A. Santaló, Integral Geometry and Geometric Probability, *Addison-Wesley, Reading, Mass.* (1976).
6. R. Schneider, Inequalities for random flats meeting a convex body, *J. Appl. Prob.* **22**, 710-716 (1985).
7. T. J. Wu, On the relations between the integrals for the power a convex body, *Acta. NanKai University.* **1** (1985).
8. G. Zhang, Dual kinematic formulas, *Trans. Amer. Math. Soc.* **351**, 895-992 (1999) .

L_P DUAL BRUNN-MINKOWSKI TYPE INEQUALITIES*

C. J. ZHAO†

Department of Information and Mathematics Sciences,
College of Science, China Institute of Metrology,
Hangzhou 310018, P.R.China
E-mail: chjzhao@163.com

G. S. LENG

Department of Mathematic,
Shanghai University,
Shanghai, P.R.China
E-mail: lenggangsong@163.com

The main purpose of this paper is to establish L_p-dual Brunn-Minkowski type inequalities, which improve three classical dual Brunn-Minkowski's type inequalities.

1 Introduction

In recent years some authors including Gardner [1-3], Lutwak [5-8], Lutwak, Yang, and Zhang [9], Schneider [10], Zhao and Leng [11-13] and *et al* have given good-sized attention to the Brunn-Minkowski theory, L^p-Brunn-Minkowski theory and their dual theories. The Brunn-Minkowski inequality was inspired by issues around the isoperimetric problem, and was for a long time considered to belong to geometry, where its significance is widely recognized. As dual form, Lutwak established three important dual Brunn-Minkowski type inequalities as follows:

Theorem 1.1 (*Dual Brunn-Minkowski inequality*, see [8]) *If* $K, L \in \mathcal{K}^n$,

*This work is supported by the academic mainstay of middle-age and youth foundation of shandong province of china.

†Work partially supported by grant 10271071 of the National Natural Science Foundation of China.

then

$$V(K\tilde{+}L)^{1/n} \leq V(K)^{1/n} + V(L)^{1/n}, \tag{1}$$

with equality if and only if K and L are dilates, where $\tilde{+}$ is the radial addition, $\mathcal{K}^n$ denote the set of all convex bodies(compact convex sets with non-empty interiors) in $\mathbb{R}^n$ and $V(K)$ denote the n-dimensional volume of convex body K.

Theorem 1.2 (*Dual Kneser-Süss inequality,* see [10]) *If $K, L \in \varphi^n$, then*

$$V(K\tilde{+}L)^{(n-1)/n} \leq V(K)^{(n-1)/n} + V(L)^{(n-1)/n}, \tag{2}$$

with equality if and only if K and L are dilates, where $\tilde{+}$ is the radial Blaschke addition, and φ^n denote the set of star bodies in $\mathbb{R}^n$.

Theorem 1.3 (see [5]) *If $K, L \in \varphi^n$, and $\lambda, \mu > 0$, then*

$$V(\lambda K\hat{+}\mu L)^{1/n} \geq \lambda V(K)^{1/n} + \mu V(L)^{1/n}, \tag{3}$$

with equality if and only if K and L are dilates, where $\hat{+}$ is the harmonic Blaschke addition.

The main purpose of the present article is to establish L_p-dual Brunn-Minkowski type inequalities, which improve above three classical inequalities, and get their generalized, reversed and strengthed forms, respectively.

2 Notation and preliminary

The setting for this paper is n-dimensional Euclidean space $\mathbb{R}^n (n > 2)$. Let $\mathbb{C}^n$ denote the set of non-empty convex figures (compact, convex subsets) and $\mathcal{K}^n$ denote the subset of $\mathbb{C}^n$ consisting of all convex bodies (compact convex subsets with non-empty interiors) in $\mathbb{R}^n$. We reserve the letter u for unit vectors, and the letter B is reserved for the unit ball centered at the origin. The surface of B is S^{n-1}.

We use $V(K)$ for the n-dimensional volume of convex body K. Let $h(K,\cdot) : S^{n-1} \to \mathbb{R}$, denote the support function of $K \in \mathcal{K}^n$; i.e., $h(K,u) = \text{Max}\{u \cdot x : x \in K\}, u \in S^{n-1}$, where $u \cdot x$ denotes the usual inner product of u and x in $\mathbb{R}^n$. The radial function $\rho(K,\cdot) : S^{n-1} \to \mathbb{R}$ of a compact subset K of $\mathbb{R}^n$ is defined by the relation $\rho(K,u) = \text{Max}\{\lambda \geq 0 : \lambda u \in K\}$. If $\rho(K,\cdot)$ is positive and continuous, K is called a star body. Let φ^n denote the set of star bodies in $\mathbb{R}^n$.

The polar coordinate formula for volume in $\mathbb{R}^n$ is

$$V(K) = \frac{1}{n}\int_{S^{n-1}} \rho(K,u)^n dS(u), \tag{4}$$

where, $dS(u)$ denotes the area element of S^{n-1} at u.

If $K_i \in \mathcal{K}^n$ $(i = 1, 2, \ldots, r)$ and λ_i $(i = 1, 2, \ldots, r)$ are nonnegative real numbers, then of fundamental importance is the fact that the volume of $\sum_{i=1}^{r} \lambda_i K_i$ is a homogeneous polynomial in λ_i given by [8]

$$V(\sum_{i=1}^{r} \lambda_i K_i) = \sum_{i_1,\ldots,i_n} \lambda_{i_1} \ldots \lambda_{i_n} V_{i_1 \ldots i_n}, \tag{5}$$

where the sum is taken over all n-tuples $(i_1, \ldots, i_n)$ of positive integers not exceeding r. The coefficient $V_{i_1 \ldots i_n}$ depends only on the bodies $K_{i_1}, \ldots, K_{i_n}$ and is uniquely determined by (5), it is called the mixed volume of $K_{i_1}, \ldots, K_{i_n}$, and is written as $V(K_{i_1}, \ldots, K_{i_n})$. Let $K_1 = \ldots = K_{n-i} = K$ and $K_{n-i+1} = \ldots = K_n = L$, then the mixed volume $V(K_1 \ldots K_n)$ is written as $V_i(K, L)$.

The radial Minkowski linear combination, $\lambda_1 K_1 \tilde{+} \cdots \tilde{+} \lambda_r K_r$, is defined by Lutwak [6]:

$$\lambda_1 K_1 \tilde{+} \cdots \tilde{+} \lambda_r K_r = \{\lambda_1 x_1 \tilde{+} \cdots \tilde{+} \lambda_r x_r : x_i \in K_i\},$$

for $K_1, \ldots, K_r \in \varphi^n$, and $\lambda_1, \ldots, \lambda_r \in \mathbb{R}$.

It has the following important property: For $K, L \in \varphi^n$ and $\lambda, \mu \geq 0$,

$$\rho(\lambda K \tilde{+} \mu L, \cdot) = \lambda \rho(K, \cdot) + \mu \rho(L, \cdot). \tag{6}$$

For $K_1, \ldots, K_r \in \varphi^n$ and $\lambda_1, \ldots, \lambda_r \geq 0$, the volume of the radial Minkowski linear combination $\lambda_1 K_1 \tilde{+} \cdots \tilde{+} \lambda_r K_r$ is a homogeneous nth polynomial in the λ_i,

$$V(\lambda_1 K_1 \tilde{+} \cdots \tilde{+} \lambda_r K_r) = \sum \tilde{V}_{i_1,\ldots,i_n} \lambda_{i_1} \cdots \lambda_{i_n}, \tag{7}$$

where the sum is taken over all n-tuples $(i_1, \ldots, i_n)$ whose entries are positive integers not exceeding r. If we require the coefficients of the polynomial in (7) to be symmetric in their argument, then they are uniquely determined. The coefficient $\tilde{V}_{i_1,\ldots,i_n}$ is nonnegative and depends only on the bodies $K_{i_1}, \ldots, K_{i_n}$. Here we denote $\tilde{V}_{i_1,\ldots,i_n}$ to $\tilde{V}(K_{i_1}, \ldots, K_{i_n})$ and is called the dual mixed volume of $K_{i_1}, \ldots, K_{i_n}$. If $K_1 = \cdots = K_{n-i} = K$, $K_{n-i+1} = \cdots = K_n = B$, the dual mixed volumes $\tilde{V}_i(K, B)$ is written as $\tilde{W}_i(K)$ and is called i-dual Quermassintegrals.

For $K_i \in \varphi^n$, then [8]

$$\tilde{W}_i(K) = \frac{1}{n} \int_{S^{n-1}} \rho(K, u)^{n-i} dS(u). \tag{8}$$

On the other hand, Let $K, L \in \varphi^n$ and $s \neq 0$, define a star body $K\tilde{+}_s L$ by

$$\rho(K\tilde{+}_s L, u)^s = \rho(K, u)^s + \rho(L, u)^s. \tag{9}$$

The operation $\tilde{+}_s$ is called p-radial addition. The radial addition $\tilde{+}$ is the special case of the p-radial addition.

Dual Quermassintegrals are special cases of the p-th dual volume:

$$\tilde{V}_p(K) = \frac{1}{n}\int_{S^{n-1}} \rho(K, u)^p dS(u), \quad -\infty < p < +\infty. \tag{10}$$

Taking $p = n$ in (10), (10) changes to (4).

A new addition, harmonic Blaschke addition, was defined by Lutwak [5]. Suppose $K, L \in \varphi^n$, and $\lambda, \mu \geq 0$(not both zero). To define the harmonic Blaschke linear combination, $\lambda K\hat{+}\mu L$, first define $\xi > 0$ by

$$\xi^{1/(n+1)} = \frac{1}{n}\int_{S^{n-1}} [\lambda V(K)^{-1}\rho(K, u)^{n+1} + \mu V(L)^{-1}\rho(L, u)^{n+1}]^{n/(n+1)} dS(u). \tag{11}$$

The body $\lambda K\hat{+}\mu L \in \varphi^n$ is defined as the body whose radial function is given by [5]

$$\xi^{-1}\rho(\lambda K\hat{+}\mu L, \cdot)^{n+1} = \lambda V(K)^{-1}\rho(K, \cdot)^{n+1} + \mu V(L)^{-1}\rho(L, \cdot)^{n+1}. \tag{12}$$

From this definition and (4), it follows immediately that $\xi = V(\lambda K\hat{+}\mu l)$, and hence

$$V(\lambda K\hat{+}\mu L)^{-1}\rho(\lambda K\hat{+}\mu L, \cdot)^{n+1} = \lambda V(K)^{-1}\rho(K, \cdot)^{n+1} + \mu V(L)^{-1}\rho(L, \cdot)^{n+1}. \tag{13}$$

3 Main results

Our main results are given in the following theorems.

Theorem 3.1 *If* $K, L \in \varphi^n$, $-\infty < s, p < +\infty$ $s \neq 0, p \neq 0$,*and* $\alpha \in [0, 1]$, *then for* $p/s > 1$

$$\tilde{V}_p(K\tilde{+}_s L)^{s/p} \leq \tilde{V}_p(\alpha K\tilde{+}_s(1-\alpha)L)^{s/p} + \tilde{V}_p((1-\alpha)K\tilde{+}_s\alpha L)^{s/p}$$

$$\leq \tilde{V}_p(K)^{s/p} + \tilde{V}_p(L)^{s/p}. \tag{14}$$

In each case, the sign of equality holds for $p \neq s$ *if and only if* K *and* L *are equivalent by dilatation.*

The inequality is REVERSED for $0 < p/s < 1$ or $p/s < 0$.

Proof From (9), (10), Minkowski integral inequality (see [8]) and notice that $\left(\int_{S^{n-1}} \rho(K,u)^p dS(u)\right)^{s/p}$ will be written $\|\rho(K,u)\|_p^s$, we obtain that for $p/s > 1$

$$\begin{aligned}\frac{\tilde{V}_p(K\tilde{+}_sL)^{s/p}}{n^{-s/p}} &= \|\rho(K\tilde{+}_sL,u)\|_p^s \\ &= \|\rho(K,u)^s + \rho(L,u)^s\|_{p/s} \\ &= \|(\alpha\rho(K,u)^s + (1-\alpha)\rho(L,u)^s) \\ &+ ((1-\alpha)\rho(K,u)^s + \alpha\rho(L,u)^s)\|_{p/s} \\ &\leq \|\alpha\rho(K,u)^s + (1-\alpha)\rho(L,u)^s\|_{p/s} \\ &+ \|(1-\alpha)\rho(K,u)^s + \alpha\rho(L,u)^s\|_{p/s} \\ &= \|\rho(\alpha K\tilde{+}_s(1-\alpha)L,u)^s\|_{p/s} + \|\rho(\alpha K\tilde{+}_s(1-\alpha)L,u)^s\|_{p/s} \\ &= \frac{1}{n^{-s/p}}(\tilde{V}_p(\alpha K\tilde{+}_s(1-\alpha)L)^{s/p} + \tilde{V}_p((1-\alpha)K\tilde{+}_s\alpha L)^{s/p}),\end{aligned}$$

On the other hand, notice that $p/s > 1$, we have

$$\tilde{V}_p(\alpha K\tilde{+}_s(1-\alpha)L)^{s/p} + \tilde{V}_p((1-\alpha)K\tilde{+}_s\alpha L)^{s/p}$$

$$\begin{aligned}&= n^{-s/p}(\|\rho(\alpha K\tilde{+}_s(1-\alpha)L,u)\|_p^s + \|\rho((1-\alpha)K\tilde{+}_s\alpha L,)\|_p^s) \\ &= n^{-s/p}(\|\alpha\rho(K,u)^s + (1-\alpha)\rho(L,u)^s\|_{p/s} \\ &+ \|(1-\alpha)\rho(K,u)^s + \alpha\rho(L,u)^s\|_{p/s}) \\ &\leq n^{-s/p}(\alpha\|\rho(K,u)\|_p^s + n^{-s/p}(1-\alpha)\|\rho(L,u)\|_p^s \\ &+ (1-\alpha)\|\rho(K,u)\|_p^s + \alpha\|\rho(L,u)\|_p^s) \\ &= n^{-s/p}\|\rho(K,u)\|_p^s + n^{-s/p}\|\rho(L,u)\|_p^s \\ &= \tilde{V}_p(K)^{s/p} + \tilde{V}_p(L)^{s/p}.\end{aligned}$$

Therefore,

$$\tilde{V}_p(\alpha K\tilde{+}_s(1-\alpha)L)^{s/p} + \tilde{V}_p((1-\alpha)K\tilde{+}_s\alpha L)^{s/p} \leq \tilde{V}_p(K)^{s/p} + \tilde{V}_p(L)^{s/p}.$$

with equality if and only if K and L are equivalent by dilatation.

In view of the inverse Minkowski integral inequality [4], similar above proof, the cases of $0 < p/s < 1$ or $p/s < 0$ easily follows. Here we omit the details.

The proof is complete.

Taking $p = n - i$ and $s = 1$ in (14), we obtain that

Corollary 3.1 *If* $K, L \in \mathcal{K}^n$ and $i < n$, *then for* $\alpha \in [0,1]$,

$$\tilde{W}_i(K\tilde{+}L)^{1/(n-i)} \leq \tilde{W}_i(\alpha K\tilde{+}(1-\alpha)L)^{1/(n-i)} + \tilde{W}_i((1-\alpha)K\tilde{+}\alpha L)^{1/(n-i)}$$

$$\leq \tilde{W}_i(K)^{1/(n-i)} + \tilde{W}_i(L)^{1/(n-i)}, \tag{15}$$

In each case, the sign of equality holds if and only if K and L are dilates.

The inequality is REVERSED for $i > n$ or $n - 1 < i < n$.

In inequality (15), let $\alpha = 0$ or $\alpha = 1$, (15) changes to classical dual Brunn-Minkowski inequality for Quermassintegrals.

Remark 3.1 Taking $p = n$ in (14), (14) changes to

$$V(K \tilde{+}_s L)^{s/n} \leq V(\alpha K \tilde{+}_s (1-\alpha)L)^{s/n} + V((1-\alpha)K \tilde{+}_s \alpha L)^{s/n}$$

$$\leq V(K)^{s/n} + V(L)^{s/n}, \tag{16}$$

Taking $\alpha = 0$ or $\alpha = 1$ and in (16), inequality (16) changes to the following result

$$V(K \tilde{+}_s L)^{s/n} \leq V(K)^{s/n} + V(L)^{s/n}, \ 0 < s \leq n \tag{17}$$

with equality for $s \neq n$ if and only if K and L are equivalent by dilatation. The reverse inequality holds when $s > n$ or when $s < 0$. This is just the s-dual Brunn-Minkowski inequality which was given by [2].

Taking $s = 1$ in (17), (17) reduces to the dual Brunn-Minkowski inequality (1) which was stated in introduction.

Taking $s = n-1$ in (17), (17) reduces to the dual Kneser-Süss inequality (2) which was stated in introduction.

Theorem 3.2 If $K, L \in \varphi^n, \lambda > 0, \mu > 0,$ *and* $-\infty < p < +\infty, \ p \neq 0,$ *then for* $p < 0$ or $0 < p < n+1$

$$\frac{\tilde{V}_p(\lambda K \hat{+} \mu L)^{(n+1)/p}}{V(\lambda K \hat{+} \mu L)} \geq \frac{\lambda \tilde{V}_p(K)^{(n+1)/p}}{V(K)} + \frac{\mu \tilde{V}_p(L)^{(n+1)/p}}{V(L)}, \tag{18}$$

with equality if and only if K and L are dilates.

The inequality is REVERSED for $p > n+1$.

Proof From (10), (13) and in view of inverse integral Minkowski inequality, we obtain that for $0 < p < n+1$ and $p < 0$

$$\begin{aligned}
\tilde{V}_p(\lambda K \hat{+} \mu L)^{(n+1)/p} &= n^{-(n+1)/p} \|\rho(\lambda K \hat{+} \mu L, u)\|_p^{n+1} \\
&= n^{-(n+1)/p} \|\xi \lambda V(K)^{-1} \rho(K,u)^{n+1} \\
&+ \xi \mu V(L)^{-1} \rho(L,u)^{n+1}\|_{p/(n+1)} \\
&\geq n^{-(n+1)/p} (\|\xi \lambda V(K)^{-1} \rho(K,u)^{n+1}\|_{p/(n+1)} \\
&+ \|\xi \mu V(L)^{-1} \rho(L,u)^{n+1}\|_{p/(n+1)}) \\
&= \xi \lambda V(K)^{-1} \tilde{V}_p(K)^{(n+1)/p} + \xi \mu V(L)^{-1} \tilde{V}_p(L)^{(n+1)/p}.
\end{aligned}$$

Dividing by $\xi = V(\lambda K \hat{+} \mu L)$, we get (18).

Remark 3.2 Taking $p = n$ in (18), inequality (18) changes to inequality (3) which was stated in introduction.

Taking $p = n - i$ in (18), (18) changes to

$$\frac{\tilde{W}_i(\lambda K \hat{+} \mu L)^{(n+1)/(n-i)}}{V(\lambda K \hat{+} \mu L)} \geq \frac{\lambda \tilde{W}_i(K)^{(n+1)/(n-i)}}{V(K)} + \frac{\mu \tilde{W}_i(L)^{(n+1)/(n-i)}}{V(L)}, \quad (19)$$

with equality if and only if K and L are dilates.

Taking $i = 0$ in (19), (19) changes to inequality (3).

Theorem 3.3 *If $K, L \in \varphi^n, \lambda > 0$, $\mu > 0$, $\alpha \in [0, 1]$ and $-\infty < p < +\infty$, $p \neq 0$, then for $p < 0$ or $0 < p < n + 1$*

$$\frac{\tilde{V}_p(\lambda K \hat{+} \mu L)^{(n+1)/p}}{V(\lambda K \hat{+} \mu L)} \geq \frac{\tilde{V}_p(\alpha \cdot \lambda K \hat{+} (1-\alpha)\mu L)^{(n+1)/p}}{V(\alpha \lambda K \hat{+} (1-\alpha)\mu L)}$$

$$+ \frac{\tilde{V}_p((1-\alpha) \cdot \lambda K \hat{+} \alpha \cdot \mu L)^{(n+1)/p}}{V((1-\alpha)\lambda K \hat{+} \alpha \mu L)}$$

$$\geq \frac{\lambda \tilde{V}_p(K)^{(n+1)/p}}{V(K)} + \frac{\mu \tilde{V}_p(L)^{(n+1)/p}}{V(L)}. \quad (20)$$

In each case, the sign of equality holds if and only if K and L are dilates.

The inequality is REVERSED for $p > n + 1$.

Proof Let $p < 0$ or $0 < p < n + 1$, we have

$$\tilde{V}_p(\lambda K \hat{+} \mu L)^{(n+1)/p} = n^{-(n+1)/p} \|\rho(\lambda K \hat{+} \mu L, u)\|_p^{n+1}$$

$$= n^{-(n+1)/p} \|\xi \lambda V(K)^{-1} \rho(K, u)^{n+1} + \xi \mu V(L)^{-1} \rho(L, u)^{n+1}\|_{p/(n+1)}$$
$$\geq n^{-(n+1)/p} \|\alpha \xi \lambda V(K)^{-1} \rho(K, u)^{n+1} + (1-\alpha) \xi \mu V(L)^{-1} \rho(L, u)^{n+1}\|_{p/(n+1)}$$
$$+ n^{-(n+1)/p} \|(1-\alpha) \xi \lambda V(K)^{-1} \rho(K, u)^{n+1} + \alpha \xi \mu V(L)^{-1} \rho(L, u)^{n+1}\|_{p/(n+1)}.$$

Hence

$$\frac{\tilde{V}_p(\lambda K \hat{+} \mu L)^{(n+1)/p}}{V(\lambda K \hat{+} \mu L)}$$

$$\geq \frac{n^{-(n+1)/p} \|\rho(\alpha \lambda K \hat{+} (1-\alpha)\mu L, u)^{n+1}\|_{p/(n+1)}}{V(\alpha \lambda K \hat{+} (1-\alpha)\mu L)}$$
$$+ \frac{n^{-(n+1)/p} \|\rho((1-\alpha)\lambda K \hat{+} \alpha \mu L, u)^{n+1}\|_{p/(n+1)}}{V((1-\alpha)\lambda K \hat{+} \alpha \mu L)}$$
$$= \frac{\tilde{V}_p(\alpha \lambda K \hat{+} (1-\alpha)\mu L)^{(n+1)/p}}{V(\alpha \lambda K \hat{+} (1-\alpha)\mu L)} + \frac{\tilde{V}_p((1-\alpha)\lambda K \hat{+} \alpha \mu L)^{(n+1)/p}}{V((1-\alpha)\lambda K \hat{+} \alpha \mu L)}.$$

On the other hand, using the inequality (18) in Theorem 3.2, we have

$$\frac{\tilde{V}_p(\alpha\lambda K\hat{+}(1-\alpha)\mu L)^{(n+1)/p}}{V(\alpha\lambda K\hat{+}(1-\alpha)\mu L)} \geq \frac{\alpha\lambda\tilde{V}_p(K)^{(n+1)/p}}{V(K)} + \frac{(1-\alpha)\mu\tilde{V}_p(L)^{(n+1)/p}}{V(L)}. \tag{21}$$

and

$$\frac{\tilde{V}_p((1-\alpha)\lambda K\hat{+}\alpha\mu L)^{(n+1)/p}}{V((1-\alpha)\lambda K\hat{+}\alpha\mu L)} \geq \frac{(1-\alpha)\lambda\tilde{V}_p(K)^{(n+1)/p}}{V(K)} + \frac{\alpha\mu\tilde{V}_p(L)^{(n+1)/p}}{V(L)}. \tag{22}$$

From (21)+(22), we have

$$\frac{\tilde{V}_p(\alpha\lambda K\hat{+}(1-\alpha)\mu L)^{(n+1)/p}}{V(\alpha\lambda K\hat{+}(1-\alpha)\mu L)} + \frac{\tilde{V}_p((1-\alpha)\lambda K\hat{+}\alpha\mu L)^{(n+1)/p}}{V((1-\alpha)\lambda K\hat{+}\alpha\mu L)}$$

$$\geq \frac{\lambda\tilde{V}_p(K)^{(n+1)/p}}{V(K)} + \frac{\mu\tilde{V}_p(L)^{(n+1)/p}}{V(L)}.$$

The proof of the cases of $p < 0$ or $0 < p < n+1$ is complete.

Similarly, the case of $p > n+1$ easily follows.

Taking $p = n - i$ in (20), (20) changed to the following result.

Corollary 3.2 If $K, L \in \varphi^n, \lambda > 0$ *and* $\mu > 0$, *then for* $n > i > -1$ or $i > n$

$$\frac{\tilde{W}_i(\lambda K\hat{+}\mu L)^{(n+1)/(n-i)}}{V(\lambda K\hat{+}\mu L)} \geq \frac{\tilde{W}_i(\alpha\cdot\lambda K\hat{+}(1-\alpha)\mu L)^{(n+1)/(n-i)}}{V(\alpha\lambda K\hat{+}(1-\alpha)\mu L)}$$

$$+\frac{\tilde{W}_i((1-\alpha)\cdot\lambda K\hat{+}\alpha\cdot\mu L)^{(n+1)/(n-i)}}{V((1-\alpha)\lambda K\hat{+}\alpha\mu L)}$$

$$\geq \frac{\lambda\tilde{W}_i(K)^{(n+1)/(n-i)}}{V(K)} + \frac{\mu\tilde{W}_i(L)^{(n+1)/(n-i)}}{V(L)}, \tag{23}$$

In each case, the sign of equality holds if and only if K and L are dilates.

The inequality is REVERSED for $i < -1$.

Remark 3.3 Taking $\alpha = 0$ or $\alpha = 1$ in (20), inequality (20) changes to inequality (18). Taking $\alpha = 0$ or $\alpha = 1$ in (23), inequality (23) changes to inequality (19).

References

1. R. J. Gardner, A positive answer to be Busemann-Petty proplem in three dimensions, *Ann. Math.*, **140**(1994), 435–477.

2. R. J. Gardner, "Geometric Tomography," Cambridge: Cambridge University Press, 1995.
3. R. J. Gardner, The Brunn-Minkowski inequality, *Bull. Amer. Math. Soc.*, **39**(2002), 355-405.
4. G. H. Hardy ,Littlewood J. E. and Pólya G. Inequalities. Cambridge Univ. Press, Cambridge, U.K., 1934.
5. E. Lutwak, Centroid bodies and dual mixed volumes, *Proc. London Math. Soc.*, **60**(1990), 365–391.
6. E. Lutwak, Intersection bodies and dual mixed volumes, *Adv. Math.*, **71**(1988), 232–261.
7. E. Lutwak, Mixed projection inequalities, *Trans. Amer. Math. Soc,* **287**(1985), 92–106.
8. E. Lutwak, Dual mixed volumes, *Pacific J.Math.*, **58**(1975), 531–538.
9. E. Lutwak, D. Yang, and G. Zhang,The Brunn-Minkowski-Firey inequality for non-convex sets, preprint.
10. R. Schneider,"Convex bodies: The Brunn-Minkowski Theory", Cambridge: Cambridge University Press, 1993.
11. C. J. Zhao and G. S. Leng, Inequalities for dual quermassintegrals of mixed intersection bodies, *Proc. Indian Acad. Sci.(Math. Sci.)*, **115**(2005), 79-91.
12. C. J. Zhao and G. S. Leng, Brunn-Minkowski inequality for mixed intersection bodies, *J. Math. Anal. Appl.*, **301**(2005), 115-123.
13. C. J. Zhao and G. S. Leng, On the polars of mixed projection bodies, *J. Math. Anal. Appl.*, 2005, accept.

ON THE RELATIONS OF A CONVEX SET AND ITS PROFILE

SHOUGUI LI AND YICHENG GONG

College of Science
Wuhan University of science and Technology
Wuhan 430081, China
E-mail: gyc4298@163.com

A compact convex set K in E^n must be the convex hull of its profile $P(K)$. However, this is not necessarily true for a non-compact convex set. When will K be the convex hull of $P(K)$? This is an open problem proposed by Steven R. Lay and unresolved up to now. The paper attempts to solve the open problem partly. It is found that when K is an open set in E^n, the answer to the problem is negative. And for an arbitrary convex set K in E^2, the paper has presented and proved a sufficient and necessary condition for the problem.

1. Introduction

Steven R. Lay postulates that if S is a nonempty compact convex set in E^n, then S must be the convex hull of its profile. However, for a non-compact convex set, this is not necessarily true. The questions then would be: when will K be the convex hull of $P(K)$? Is there a unique smallest subset whose convex hull equals K? What if K is an open set? What if K is a closed, a bounded or a compact set? These were first asked by Steven R.Lay as an open problem .[1] This paper attempts to partly resolve these open problem. For this purpose, some notations and a definition for the profile of a convex set are given first. Let E^n, L and L_x be the n-dimensional Euclidean space, an arbitrary line in E^2 and a line through the point x in E^2 respectively. Obviously L is a hyperplane in E^2, and there must exist a linear functional F and a real number a such that $L{=}(x,y)|F(x,y) = a$. For the two open semi-planes bounded by the line L, the one $(x,y)|F(x,y) - a > 0$ is denoted by L^+, and and the other $(x,y)|F(x,y) - a < 0$ by L^-. For a convex set K, its boundary, relative interior and interior will be denoted by ∂K, *relint*(K) and $int(K)$ in order. [1] If K is an arbitrary set in E^n, we will denote its convex hull and closure by $Conv(K)$ and $Cl(K)$ respectively. For any point

x in E^n and a positive constant ε, let $B(x, \varepsilon)$ be an open ball with center x and radius ε.

Definition 1 A point P of K is called an extreme point of K if there is not any non-degenerate line segment in K such that P lies in the relative interior of the segment. The set consisting of all the extreme points of K is called the profile of K and denoted by $P(K)$.

To resolve the problem, we propose here a new concept called the minimal generating subset.

Definition 2 For a nonempty convex set K and a subset $M \subset K$, M is called a minimal generating subset of K, if and only if:
(1) $\forall x \in K$, x can be expressed as a combination of finite points of M, namely $K \subset Conv(M)$.
(2) $\forall y_i \in M$, y_i can not be expressed as a combination of finite points of $M - \{y_i\}$, namely $y_i \notin Conv(M/\{y_i\})$.

Remark 1: The first condition suggests $M \subset K$ and M can convexly generate K, which, we know by *Caratheodory* Theorem ,[2] can be easily satisfied.
The second condition suggests that M is a convexly independent system. So they are counter-parts in a linearly independent maximal subset of a linear space. It is also worthy of attention, that in terms of functions, what a minimal generating subset to a convex set are what a maximal linearly independent subset to a linear space. It is well known that the maximal linearly independent subset of a linear space of finite dimensions must exist. However this is not necessarily true of the minimal generating subset of a convex set.

2. Main results

Based on the ideas in Ref.3-9, this paper is an attempt to partly resolve the aforementioned open problem. First, a negative answer to the question is presented when K is an open set in E^n, which is expressed in Theorem 1. Then it's proved that if K is the convex hull of $P(K)$, $P(K)$ must be the unique minimal convex generating subset of K, which leads to Theorem

2. To obtain a sufficient and necessary condition for the problem, we have prepared two lemmas and decomposed it into a sufficient condition Theorem 3 and a necessary condition Theorem 4, respectively. Combining Theorem 3 and Theorem 4, we have Theorem 5.

Theorem 1: If K is a nonempty open set in E^n, the minimal convex generating subsets of K do not exist.

Corollary 1: Suppose K is a nonempty convex set in E^n. If there is a line L such that $L^+ \cap K$ or $L^- \cap K$ is an open set, then the minimal generating sets of K do not exist.

Lemma 1: In E^n, an extreme point of a convex set must be a boundary point, that is, $p(K) \subset \partial K$.

Lemma 2: Suppose K is a nonempty convex set in E^n. Then for any point $x \in p(K) \Rightarrow x \notin Conv(K - \{x\})$. In other words, an extreme point can be expressed only in its own convex combination.

Theorem 2: If $Conv(P(K)) = K$, then $P(K)$ must be the unique minimal convex generating subset of K.

Theorem 3: If the following two conditions are satisfied at the same time, $P(K)$ is the minimal generating set of a planar convex set K.

(1)Suppose K is a nonempty planar convex set. If for any given supporting line H, $H \cap \partial K$ is a bounded set and $H \cap \partial K \cap K$ is closed in $H \cap \partial K$;

(2)For any given line $L \subset E^2$, neither $L^+ \cap K$ nor $L^- \cap K$ is an open set.

Theorem 4: If $P(K)$ is the minimal generating set of a planar convex set K, the following must both hold.

(1) For any given supporting line H, $H \cap \partial K$ is a bounded set and $H \cap \partial K \cap K$ is closed in ∂K;

(2) For any given line $L \subset E^2$, neither $L^+ \cap K$ nor $L^- \cap K$ is an open set.

Theorem 5: A sufficient and necessary condition for $P(K)$ being the minimal generating set of a planar convex set K is: for any given supporting line H, $H \cap \partial K$ is a bounded set and $H \cap \partial K \cap K$ is closed in ∂K; for any given line $L \subset E^2$, neither $L^+ \cap K$ nor $L^- \cap K$ is an open set.

3. The proof of the main results

Proof of Theorem 1: Suppose the minimal convex generating subsets of K exist,which is called M. We will take three steps to prove its absurdity.

Step 1 M can not be empty. It goes without proof.

Step 2 $M \neq K$

Step 3 M cannot be any other set.

Therefore the above assumption is absurd.

Proof of Step 2 If $M = K$ where K is an nonempty open set in E^n, then for any point $x \in M = K$, there must be a positive constant δ such that $B(x, \delta) \subset K = M$ and $x \in Conv(B(x, \delta) - \{x\}) \subset Conv(M - \{x\})$. This contradicts the fact that M is the minimal generating set of K. Therefore $M \neq K$.

Proof of Step 3 Because of the complexity of the process the proof will be decomposed into two parts.

Part 1 From the above proof we know $K - M = \phi$.We can select a fixed point $x_0 \in K - M$, then for any ray L_{x_0} with an end point x_0, $L_{x_0} \cap M$ is either an empty set or a set containing only one point; otherwise there are at least two points in $L_{x_0} \cap M$, called x_1, x_2. Without loss of the generality we can suppose $d(x_1, x_0) > x_2, x_0$. It follows that x_1 is a point of the line segment between x_0, x_2.

So there must be a constant $0 < a < 1$ such that

$$x_1 = ax_0 + (1 - a)x_2. \tag{1}$$

As $x_0 \in K$, $x_0 \in Conv(M)$. From *Caratheodory* theorem we know there are $n + 1$ points $y_1, y_2, y_3, \cdots, y_{n+1} \in M$, such that

$$x_0 = a_1y_1 + a_2y_2 + \cdots + a_{n+1}y_{n+1}, \tag{2}$$

where

$$a_i \geq 0, a_1 + a_2 + \cdots + a_m = 1. \tag{3}$$

So,

$$x_1 = a(a_1y_1 + a_2y_2 + a_3y_3 + \cdots + a_{n+1}y_{n+1}) + (1 - a)x_2. \tag{4}$$

If none of the $n + 1$ points $y_1, y_2, y_3, \cdots, y_{n+1}$ equals x_1, then we can prove the expression (4) is a convex combination expression of x_1, that is, $x_1 \in Conv(M - \{x_1\})$. This contradicts the fact that M is a minimal convex generating subset.

If one of the $n + 1$ points $y_1, y_2, y_3, \cdots, y_{n+1}$ equals x_1, for example $y_1 = x_1$, then

$$x_0 = a_1x_1 + a_2y_2 + a_3y_3 + \cdots + a_{n+1}y_{n+1}, \tag{5}$$

where

$$a_i \geq 0, a_1 + a_2 + \cdots + a_m = 1, \tag{6}$$

$$x_1 = a(a_1x_1 + a_2y_2 + a_3y_3 + \cdots + a_my_m) + (1-a)x_2. \tag{7}$$

That is,

$$(1 - aa_1)x_1 = (1-a)x_2 + aa_2y_2 + \cdots + aa_{n+1}y_{n+1}. \tag{8}$$

Obviously, $a_1 \neq 1$ is valid; otherwise there will be a contradiction between $x_0 = x_1$ and $d(x_0, x_1) > 0$

So $0 < (1 - aa_1) < 1$, and

$$x_1 = [1/(1-aa_1)] \times [(1-a)x_2 + aa_2y_2 + aa_3y_3 + \cdots + aa_my_m] \tag{9}$$

On the other hand, we know

$$\begin{aligned} &[1/(1-aa_1)] \times [(1-a) + aa_2 + aa_3 + \cdots + aa_m] \\ &= [1/(1-aa_1)] \times [(1-a) + a(1-a_1)] \\ &= [1/(1-aa_1)] \times (1-aa_1) = 1 \end{aligned} \tag{10}$$

and $(1-aa_1), (1-a), aa_2, aa_3, \cdots, aa_m$ are all positive. Therefore Eq.9 means $x_1 \in Co(M - \{x_1\})$. This contradicts the fact that M is the minimal generating set. So for any ray l_{x_0} with an endpoint x_0, $l_{x_0} \subset M$ is either an empty set or a set containing only one point.

(3) For any point $p_1 \in M \subset K$, as K is an open set, there must be a point p_2 lying in the production of the ray x_0p_1. This indicates p_1 lies in the interior of line segment x_0p_2.Thus p_1 must can be expressed as a convex combination of x_0, p_2.

However p_2 does not belong to M; otherwise the intersection of M and the ray x_0p_1 contains two points p_1, p_2. And this contradicts the conclusion reached in Part 1.

As M is a generating set, from *Carathoeory* Theorem it can be inferred that x_0, p_2 can both be expressed as a convex combination by $n+1$ or fewer points of M.

Suppose their convex combination expressions are

$$x_0 = b_1q_1 + b_2q_2 + b_3q_3 + b_{n+1}q_{n+1}, \tag{11}$$

and

$$p_1 = 1/[1 - cb_1 - (1-c)\beta_1]$$

$$\times[c(b_2 + b_3 + \cdots + b_{n+1}) + (1-c)(\beta_2 + \beta_3 + \cdots + \beta_{n+1})] \quad (12)$$

where

$$b_1 + b_2 + \cdots + b_{n+1} = 1, b_i \geq 0, q_i \in M, 1 \leq i \leq n+1. \quad (13)$$

Since $x_0 \in K - M$ and $q_1 \in M$, therefore $0 < b_1 < 1$ and

$$p_2 = \beta_1 p_1 + \beta_2 z_2 + \beta_3 z_3 + \cdots + \beta_{n+1} z_{n+1} \quad (14)$$

where

$$\beta_1 + \beta_2 + \beta_3 + \cdots + \beta_{n+1} = 1, \beta_i \geq 0, z_i \in M, 1 \leq i \leq n+1, \quad (15)$$

Since $p_2 \notin M, p_1 \in M$, so

$$\beta_1 \neq 1 \quad (16)$$

$$\begin{aligned} p_1 &= cx_0 + (1-c)p_2 \\ &= c(b_1 p_1 + b_2 p_2 + b_3 p_3 + \cdots + b_{n+1} p_{n+1}) \\ &\quad +(1-c)(\beta_1 p_1 + \beta_2 z_2 + \beta_3 z_3 + \cdots + \beta_{n+1} z_{n+1}) \\ &= [cb_1 + (1-c)\beta_1]p_1 + cb_2 q_2 + \cdots + cb_{n+1} q_{n+1} \\ &\quad +(1-c)\beta_2 z_2 + \cdots + (1-c)\beta_{n+1} z_{n+1} \end{aligned} \quad (17)$$

where $0 \leq c \leq 1$.

As $x_0 \in K - M, p_1 \in M, 0 < c < 1$.

As a result $cb_1 < 1$ and $(1-c)\beta_1 < (1-c)$. Therefore

$$1 - cb_1 - (1-c)\beta_1 > 1 - c - (1-c) = 0 \quad (18)$$

$$p_1 = 1/[1 - cb_1 - (1-c)\beta_1] \quad (19)$$

$$\times[c(b_2 + b_3 + \cdots + b_{n+1}) + (1-c)(\beta_2 + \beta_3 + \cdots + \cdots + \beta_{n+1})]$$

$$\begin{aligned} &c(b_2 + b_3 + \cdots + b_{n+1}) + (1-c)(\beta_2 + \beta_3 + \cdots + \beta_{n+1}) \\ &= c(1 - b_1) + (1-c)(1-\beta_1) = 1 - cb_1 - (1-c)\beta_1 \end{aligned}$$

and $cb_i \leq 0,\ (1-c)\beta_i \geq 0,\ 2 \leq i \leq n+1$.

So Eq.12 indicates $p_1 \in Conv(M - \{p_1\})$, and this contradicts the fact that M is a minimal convex generating subset. It's thus proved that the minimal convex generating subsets of an open set do not exist.

Proof of Corollary 1: Suppose K is a nonempty convex set in E^2. If there is a line L which cuts K into $L^+ \cap K$ and $L^- \cap K$ such that $L^+ \cap K$ (or$L^- \cap K$) is an open set.By Theorem 1 the minimal convex generating subsets of $L^+ \cap K$ (or$L^- \cap K$) do not exist. Therefore, then the minimal convex generating subsets of K do not exist.

Proof of Lemma 1: For any point $x \in P(K) \Rightarrow x \in \partial K \cup \text{int(K)}$, if $x \in \text{int(K)}$, there is a $\delta > 0$, such that the closed ball $B(x, \delta) \subset \text{int(K)}$. Therefore, there must exist a non-degenerate line segment containing x in its relative interior. As a result $x \notin P(K)$. This contradicts the premise. So $x \in \partial(K)$ and $p(K) \subset \partial K$.

Proof of Lemma 2: Suppose there is a point $x_0 \in P(K) \subset Conv(K - \{x_0\})$. By *Caratheodory* Theorem there are $n+1$ points $y_1, y_2, \cdots, y_{n+1} \in (K - \{x_0\})$ such that

$$x_0 = a_1 y_1 + a_2 y_2 + \cdots + a_{n+1} y_{n+1} \tag{20}$$

where

$$a_1 + a_2 + \cdots + a_{n+1} = 1 \tag{21}$$

$$a_i \geq 0, \ 1 \leq i \leq n+1 \tag{22}$$

with at least $1 > a_k > 0$, $1 > a_l > 0$, otherwise there must be a m such that $a_m = 1$ and $x_0 = y_m$. As a result $x_0 \in (K - \{x_0\})$, which is a contradiction. Suppose $k \leq i, i \geq i+1$ then

$$0 < [1/(a_1 + \cdots + a_i)] < 1 \tag{23}$$

$$0 < [1/(a_{i+1} + \cdots + a_{n+1})] < 1 \tag{24}$$

$$x_0 = \lambda u + (1 - \lambda) v \tag{25}$$

where

$$\lambda = a_1 + \cdots + a_i, 1 - \lambda = a_{i+1} + \cdots + a_{n+1} \tag{26}$$

$$u = [1/(a_1 + \cdots + a_i)](a_1 y_1 + \cdots + a_i y_i) \tag{27}$$

$$v = [1/(a_{i+1} + \cdots + a_{n+1})](a_{i+1} y_{i+1} + \cdots + a_{n+1} y_{n+1}) \tag{28}$$

Since K is a nonempty convex set it is easy to prove that both u and v belong to K and can be expressed as a convex combination by u, v. Therefore, x lies in the relative interior of non-degenerate line segment uv, which contradicts the fact that M is the minimal generating set of K. That is to say, the hypothesis is absurd and the prime proposition is valid.

Proof of Theorem2: Suppose there is another minimal convex generating subset S_1 of K. By **Lemma 2**, $P(K) \subset S_1$ is valid; so there is $x_1 \in S_1 - P_(K)$. As $Conv(P(K)) = K$, $x_1 \in Conv(P(K))$ which contradicts the second item of the definition of a minimal convex generating subset of K. Therefore $P(K)$ must be the unique minimal convex generating subset K.

Remark2: From Theorem 2 we can understand why the minimal convex generating subset of a bounded closed set in finite dimensional Euclidean space is the very profile.

Proof of Theorem 3: We will take two steps to prove **Theorem 3**, namely $Conv(P(K)) = K$.

In Step 1, we will prove $K \cap \partial K \subset Conv(P(K))$ as follows.

For any given $x \in K \cap \partial K$, there is $x \in (K \cap \partial K \cap P(K)) \cup (K \cap \partial K - P(K))$. Suppose $x \in K \cap \partial K \cap P(K)$. From **Lemma 1** , we know $x \in Conv(P(K))$ and the proposition is proved to be valid. Suppose, further,$x \in K \cap \partial K \cap P(K)$, that is, x is not an extreme point but in ∂K.So x is contained in the intersection of supporting line H at point x and $\partial K \cap K$. Since $H \cap \partial K$ is a bounded set, so $H \cap \partial K$ is a line segment called MN. On the other hand, $H \cap \partial K \cap K = MN \cap K$ is a line segment ab. As for the given supporting line H $x \in K \cap \partial K \cap P(K)$ is closed in ∂K, the end point must both lie in line segment $x \in K \cap \partial K \cap P(K)$. As a result

$$a \in P(K), b \in P(K) \tag{29}$$

and

$$x = \beta a + (1 - \beta)b, 0 < \beta < 1. \tag{30}$$

The proposition is true.

In Step 2, we will prove $\text{int}(K) \subset Conv(P(K))$.

It can be demonstrated that there are at least two different supporting lines $H_1, H_2$of K such that K is contained in the area between H_1 and H_2.

If K has only one or no supporting line, K must be a half-plane or plane and there must be a line $L \subset E^2$ such that $L^+ \cap K$ or $L^- \cap K$ is an open set.From **Corollary 1** it can be inferred that the minimal generating set of K doesn't exist, which contradicts the premise of the **Theorem 3**.

For any given $x \in \text{int(K)}$, there is a $\varepsilon > 0$ such that

$$B(x, \varepsilon) \subset \text{int(K)}. \tag{31}$$

In the supporting lines H_1, H_2, there must be three points which are not in a common line, which might be called A, B, C such that

$$B(x, \varepsilon) \subset Conv(A, B, C). \tag{32}$$

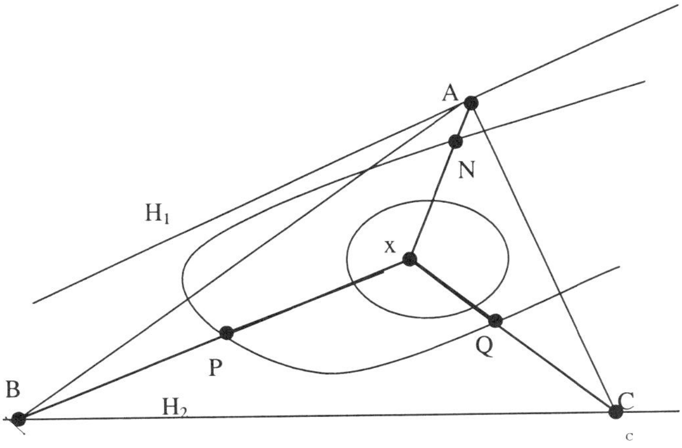

Fig. 1.

Connect x and A, B, C, respectively, then all the line segments xA, xB, xC must intersect with its boundary. Denote the intersection points by N, P, Q respectively(see Fig. 1). Take x as the original point of the coordinate system, there must be $\mu_1, \mu_2, \mu_3 \geq 0$ and $\lambda_1, \lambda_2, \lambda_3 \geq 0$ such that

$$N = \mu_1 A, P = \mu_2 B, Q = \mu_3 C \tag{33}$$

and

$$\lambda_1 A + \lambda_2 B + \lambda_3 C = 0 \tag{34}$$

$$\lambda_1 + \lambda_2 + \lambda_3 = 1 \tag{35}$$

where none of μ_1, μ_2, μ_3 is zero and none of $\lambda_1, \lambda_2, \lambda_3$ is zero.Otherwise it follows that x is a boundary point, which contradicts the fact that x is an interior point.

As a result,

$$\frac{1}{\lambda}\left[\frac{\lambda_1}{\mu_1}A + \frac{\lambda_2}{\mu_2}B + \frac{\lambda_3}{\mu_3}C\right] = 0 \tag{36}$$

where

$$\lambda = \frac{\lambda_1}{\mu_1} + \frac{\lambda_2}{\mu_2} + \frac{\lambda_3}{\mu_3}. \tag{37}$$

This expression can be easily proved to be a convex combination expression of the original point(if x is not the original point we need only to make a translation). As a result, $x \in Conv(N, P, Q)$.

(i) If all of N, P, Q belong to $\partial K \cap K$ and from Step 1 we have

$$N, P, Q \in Conv(P(K)). \tag{38}$$

Then $x \in Conv(P(K))$ and the conclusion holds.

(ii) If $N \notin \partial K \cap K$ and $P, Q \in \partial K \cap K$ then N is either an exposure point or lying in the relative interior of the line segment uv. If N is an exposure point, we can turn xN a small angle centered at x in the direction clockwise and counter-clockwise respectively, such that it can intersect ∂K at points N_1, N_2. N_1, N_2 must both belong to $\partial K \cap K$;otherwise the line NN_1 (or NN_2) will result in an open set out of $L^+ \cap K$ or $L^- \cap K$(E.g. line NN_1 (or NN_2) and arc $NN_1(NN_2)$ form an open subset of K), which contradict the premise of **Theorem 3**. From **the first step** we have $N_1, N_2, P, Q \in Conv(P(K))$. Since $x \in Conv(N_1, N_2, P, Q)$, so $x \in Conv(P(K))$.

If N lies in the relative interior of the line segment uv and $P \notin K \cap \partial K$, at least one of the end points u, v can not belong to $\partial K \cap K$. Suppose u does not belong to $\partial K \cap K$, and there must be a point u_1 of $\partial K \cap K$ in the side that is near enough to the end point u and opposite to the line segment uv; otherwise there would be a line $L \subset E^2$ such that $L^+ \cap K$ or $L^- \cap K$ is an open set. From Corollary 1it's known that this contradicts the premise of the proposition.

If v does not belong to $\partial K \cap K$, the problem can be dealt with similarly. We can find a point x that belongs to $\partial K \cap K$ such that $B(x, \varepsilon) \subset Conv(u_1, v_1, P, Q)$ and so $x \in Conv(P(K))$.

(iii) If $N \notin \partial K \cap K$ and $Q, P \in \partial K \cap K$, $P \notin \partial K \cap K$ and $Q, N \in \partial K \cap K$, similar steps can be taken.

(iv) Suppose $Q \notin \partial K \cap K$ or $N \notin \partial K \cap K$ or $P \notin \partial K \cap K$.we can deal the exposure point or the point in the interior of a line segment using aforementioned method respectively.In other words, in this condition $x \in Conv(P(K))$ still holds.

By now we have proved $Conv(P(K)) \supset K$. Since K is a convex set so $Conv(P(K)) = K$ and by Lemma 2 we know $P(K)$ is K's unique minimal convex generating subset.

Proof of Theorem 4: The theorem suggests, if $P(K)$ is the minimal generating set of a planar convex set K, the 3 conclusions must hold. We will cut the proof into three parts correspondingly.

(1) For any given line $L \subset E^2$, neither $L^+ \cap K$ nor $L^- \cap K$ is an open set. This is the inverse and negative proposition of Corollary 1. So it must hold.

(2) If $H \cap \partial K$ is a unbounded set, we might as well suppose $H \cap \partial K$ is a ray L_1 whose end point is A.

Then $L_1 \cap K$ can be but an empty set or a set containing only one point or line segment or ray. However we will prove none of these four cases can hold.

(i) Suppose $L_1 \cap K$ is an empty set. Select a point $P \in L_1 \cap K$, we can draw a line L_p through P such that $L_1 \cap L_p = \{P\}$ while the angle between L_1 and L_p small enough such that $L_p \cap \partial K \cap K$ is an empty set. Then at least one of $L_p^+ \cap K$ and $L_p^- \cap K$ would be an angle area. That is, at least one would be an open set. And by Corollary 1 the minimal generating set of K would not exist which contradicts the premise of **this Theorem**. So $L_1 \cap K$ must be nonempty.

(ii) Suppose $L_1 \cap K$ is a set containing only one point, which is called Q. Select a point $P \in L_1 \cap \partial K$ such that $d(Q, A) < d(P, A)$. And the intersection of K and L_2, which is a ray with an end point P in the ray L_1, would be an empty set. And this, by (i), is impossible.

(iii) Suppose $L_1 \cap K$ is a line segment whose end points are M_1, M_2. Select $p \in L \cap \partial K$ such that $d(P, A) > d(P, M_1)$ and $d(P, M) > d(P, M_2)$, then P would not belong to $L \cap \partial K \cap K$ and the intersection of K and L_p, which is a ray with an end point p in the ray L_1, would be an empty set.And this, by (i), is impossible, too.

(iv) Suppose $L_1 \cap K$ is a ray, which is called L_2 with an end point B. Then L_2 must be able to be generated by $P(K)$ convexly.

Since L_2 is a ray, L_2 can be generated by at least three points, might as well called as X_1, X_2, X_3. Suppose $d(X_1, X_2) < d(X_1, X_3)$, and then

$X_2 \in rel\text{int}(X_1, X_3)$, that is, X_2 can be expressed as a convex combination of X_1, X_3. Therefore $Conv(P(K) - \{X_2\}) = K$ which contradicts the premise that $P(K)$ is the minimal generating set.

In a word, the hypothesis that $H \cap \partial K$ is an un-bounded set is absurd and so it should be bounded. The second part of the theorem holds.

(3) For any given supporting line H, by (2), $H \cap \partial K \cap K$ is a bounded line segment ab. We will prove $H \cap \partial K \cap K$ is closed in ∂K. Otherwise $a \notin H \cap \partial K \cap K$ or $b \notin H \cap \partial K \cap K$, and it follows readily that line segment ab can only be generated by intersection of $P(K)$ and line segment ab. Since the minimal generating set of non-closed line segment ab contains at least two points b and x (here $a \notin H \cap \partial K \cap K$), where $x \in$ line segment ab.

Since on the other hand an arbitrary point $y \in$ line segment ax, $y \notin Conv(x, b)$, a set of two points can not be the minimal generating set of line segment ab. If the minimal generating set of non-closed line segment contains three points x_1, x_2, x_3 , supposing $d(x_1, x_2) < d(x_1, x_3)$,and it would follow that $x_2 \in rel\text{int}(x_1, x_3)$, so $x_2 \in Conv(x_1, x_3)$. Therefore the minimal generating set of non-closed line segment ab does not exist. Further more the line segment ab must be a closed line segment, that is, a, b are both contained in $H \cap \partial K \cap K$.So $H \cap \partial K \cap K$ is closed in ∂K.

As the three parts of the theorem have been proved, the whole theorem must hold.

Combining Theorem 3 and Theorem 4, we will arrive at a sufficient and necessary condition for the open problem, namely **Theorem 5**.

4. Conclusion

The conclusion of **Theorem 5** is significant. It not only covers the conclusion in the Ref. 1 that the minimal convex generating subset of a planar compact convex set must be its profile, but also states that there are some planar non-compact convex sets which meet the conditions of **Theorem 5** and whose minimal convex generating subset does exist. For example a bounded convex set with only finite boundary points not in the set is not a compact convex set but satisfies the conditions of **Theorem 5** and the conclusion $Conv(P(K)) = K$ still holds.

References

1. S. Lay, Convex Sets and Their Applications, New York: A Willey-interscience publication B, (1972).

2. R. Schneider, Convex bodies: the Brunn Minkowski theory, New York: Cambrige University Press, (1993).
3. R. J. Gardner, Geometric tomography, New York: Cambrige University Press, (1995)
4. M. Berger, Convexity, Amer. Math. Mothly, 97(1990), 650-678.
5. R. Burton, Sections of convex Bodies, J. London math. Soc., (2)12(1976), 331-336.
6. K. Ball, Some remarks on the geometry of convex sets, In: Geometric aspects of Functional Analysis (J. Lindenstrauss and V.D. Milman, Eds), Lecture Notes in Math., vol. 1317, Springer, Berlin, (1988) 224-231.
7. Ren Delin, Topics in Integral Geometry, World Scientific, (1994).
8. Ren Delin, Two topics in integral geometry, Differential Geometry and Differential Equations, Proc. Symp., Shanghai, (1984), 309-333.
9. J. Zhou, The sufficient condition for a convex body to fit another in R^4, Proc. Amer. Math. Soc., 121(1994), 907-913.

CONVEX BODIES WITH SYMMETRIC X-RAYS IN TWO DIRECTIONS

LI DEYI

College of science, Wuhan university of science and technology, Wuhan, 430081, China,
E-mail: lideyiwkd@yahoo.com.cn

XIONG GE

Department of Mathematics, Shanghai university, Shanghai, 200444, P.R.China,
E-mail: xiongge@staff.shu.edu.cn

In this paper, we obtain and prove two sufficient conditions. Firstly, we give a sufficient condition for a convex body being origin-symmetric. Secondly, we obtain a sufficient condition for a convex body with constant width being a disc.

1. Basic concepts and notation

The origin is denoted by o and the unit circle in E^2 by S^1. A direction is identified with the corresponding unit vector u, a point in S^1. The line through o parallel to u is denoted by l_u, and the line through o orthogonal to u by $u^\perp$.

The $X-ray$ X_uK [1] of a planar convex body K in the direction u is a function, which defined on $u^\perp$, giving the length of each chord of K parallel to u. $K|u^\perp$ is the projection of K in $u^\perp$. Obviously, the support of X_uK, which is the closure of point set $\{x \in u^\perp | X_uK(x) \neq 0\}$, is $K|u^\perp$.

Suppose $u \in S^1$, the *Steiner symmetral* S_uK [1] of K in the direction u is the set obtained by translating, in the direction u, all the chords of K parallel to u so that they are bisected by $u^\perp$, and then taking the union of the resulting line segments. The boundary of K is denoted by bd$K$ and the area of K by $\lambda_2(K)$. The support function h_K of K is defined by $h_K(u) = max\{z \cdot u | z \in K\}$[1][2]. The width function $w_K(u)$ of K can be expressed as $w_K(u) = h_K(u) + h_K(-u)$ [1][2]. F_u is a map defined in bd$K$ that interchanges the endpoints of the chords of K parallel to u [1], where

K is a strictly convex body. When the length of chord is equal to 0, the chord degenerates into a point P, so $F_u(P) = P$.

Now set up the perpendicular coordinate system by $u^{\perp}$ as the $x-$ axis and l_u as $y-$axis. Suppose $K = \{(x,y)| -a \le x \le a, g(x) \le y \le f(x)\}$, where $f(x)$ is a concave function and $g(x)$ is a convex function, and $X_uK(-x) = X_uK(x)$. For $A(x, f(x)) \in \mathrm{bd}K$ (or $B(x, g(x)) \in \mathrm{bd}K$), $V_u(A) = \{A(x, f(x)), B(x, g(x)), C(-x, f(-x)), D(-x, g(-x))\}$ is called the vertex set associated with A (or B) in the direction u. It is obvious that the quadrangle $ABCD$ is a parallelogram because of $X_uK(-x) = X_uK(x)$. We denote such a parallelogram by $G_u(A)$ and call it the parallelogram associated with A in direction u. When $A \in l_u \bigcap \mathrm{bd}K$, $G_u(A)$ degenerates into a line segment. Assume the centroid of K is $W_K(\xi_K, \eta_K)$. Let Θ_u be the set of point P on $\mathrm{bd}K$, where the centroid of $G_u(P)$ is $W_K(\xi_K, \eta_K)$.

2. Lemmas

In the following, we use v, u to be the unit vector in directions of $x-$axis and $y-$axis ,respectively.

Lemma 2.1. *Let $K = \{(x,y)| -a \le x \le a, g(x) \le y \le f(x)\}$ be a convex body in E^2 and $X_uK(-x) = X_uK(x)$. Then*

$$\xi_K = 0 \tag{1}$$

$$\eta_K = \frac{1}{2\lambda_2(K)} \int_0^a X_uK(x)(f(x) + g(x) + f(-x) + g(-x))dx \tag{2}$$

Proof. $\xi_K = 0$ is easy.

$$\begin{aligned}
\eta_K &= \frac{1}{\lambda_2(K)} \iint\limits_K y\mathrm{d}x\mathrm{d}y \\
&= \frac{1}{\lambda_2(K)} \int_{-a}^{a} \mathrm{d}x \int_{g(x)}^{f(x)} y\mathrm{d}y \\
&= \frac{1}{\lambda_2(K)} \int_{-a}^{a} \frac{1}{2}(f^2(x) - g^2(x))\mathrm{d}x \\
&= \frac{1}{2\lambda_2(K)} \int_{-a}^{a} X_uK(x)(f(x) + g(x))\mathrm{d}x \\
&= \frac{1}{2\lambda_2(K)} \int_{0}^{a} X_uK(x)(f(x) + g(x) + f(-x) + g(-x))\mathrm{d}x
\end{aligned}$$

□

Lemma 2.2. *Let $K = \{(x, y)| -a \le x \le a, g(x) \le y \le f(x)\}$ be a strictly convex body in E^2 and $X_uK(-x) = X_uK(x)$. Then Θ_u is a closed set.*

Proof. If Θ is empty, Θ is certainly closed.

If Θ is not empty, assume $A_0(x_0, y_0) \in \overline{\Theta}_u$, then there exist $A_n(x_n, y_n) \in \Theta_u$ such that $A_n(x_n, y_n) \to A_0(x_0, y_0)$. For each n, either $y_n = f(x_n)$ or $y_n = g(x_n)$. So there are infinite n_i such that $y_{n_i} = f(x_{n_i})$ or there are infinite n_i such that $y_{n_i} = g(x_{n_i})$. Without loss of generality, suppose $y_n = f(x_n)$ holds for all n. Let the vertexes of the parallelogram associated with $A_n(x_n, y_n)$ in direction u be $A_n(x_n, f(x_n))$, $B_n(-x_n, f(-x_n))$, $C_n(-x_n, g(-x_n))$, $D_n(x_n, g(x_n))$. Since $f(x)$ and $g(x)$ are continuous and $x_n \to x_0$, then $f(x_n) \to f(x_0)$, $g(x_n) \to g(x_0)$. so $A_n \to A_0$, $B_n \to B_0$, $C_n \to C_0$, $D_n \to D_0$. The quadrangle $A_0B_0C_0D_0$ is the parallelogram associated with $A_0(x_0, y_0)$ in direction u.
Since $A_n \in \Theta_u$, so

$$\frac{1}{4}[f(x_n) + g(x_n) + f(-x_n) + g(-x_n)] = \eta_K \tag{3}$$

Let $n \to \infty$, then

$$\frac{1}{4}[f(x_0) + g(x_0) + f(-x_0) + g(-x_0)] = \eta_K \tag{4}$$

So the centroid of the parallelogram $A_0B_0C_0D_0$ coincide with the centroid $W_K(0, \eta_K)$ of K. Hence $A_0 \in \Theta_u$ and therefore Θ_u is closed.

Let $\Theta = \Theta_u \cup \Theta_v$, then Θ is a closed set. □

Lemma 2.3. *Let K be a strictly convex body in E^2. Suppose X_uK and X_vK are both even, then Θ is origin-symmetric*

Proof. Since $X_uK(x)$ and $X_vK(y)$ are both even, so the centroid of K is at the origin. For arbitrary $A \in \Theta$ there exist a parallelogram $G(A)$ with centroid at origin and with A as one of vertex. So the point C opposite to A is on bd$K$, and A is symmetric with C about origin. Hence Θ is an origin- symmetric set.

□

Lemma 2.4. *Let K be a strictly convex body in E^2. Suppose X_uK and X_vK are both even and $X_uK(0) = w_K(u)$, $X_vK(0) = w_K(v)$. For each point $P \in$ bd$K$, (i) if $F_u(P) \in \Theta$ or $F_v(P) \in \Theta$, then $P \in \Theta$. (ii) if $P \in \Theta$, then $F_u(P) \in \Theta$ and $F_v(P) \in \Theta$.*

Proof. Since $X_uK(x)$ and $X_vK(y)$ are both even, so their support sets are symmetric interval denoted by $[-a,a]$ and $[-b,b]$, respectively. Therefore the centroid of K is at the origin o, and $w_K(u) = 2b$, $w_K(v) = 2a$. From the facts $X_uK(0) = w_K(u) = 2b$, $X_vK(0) = w_K(v) = 2a$, the intersection points of bd$K$ with coordinate axes are $(-a,0),(a,0)$,$(0,-b)$,$(0,b)$ and bd$K$ is divided into four parts belonging to four quadrant respectively. Since lines $x = -a, x = a, y = -b, y = b$ are all the support lines of K, so the two endpoints of arbitrary level chord are situated at the two sides of $y-$axis and the two endpoints of arbitrary vertical chord are situated at the two sides of $x-$axis.

(i) Suppose $P \in$ bd$K$ and $F_v(P) \in \Theta = \Theta_u \cup \Theta_v$, let $A = F_v(P)$. Without loss of generality, Assume A is situated at the first quadrant. It is obvious that $G_v(A) = G_v(P)$.

1. If $A \in \Theta_v$, then the centroid of $G_v(A)$ is at the origin o. It follows that the centroid of the parallelogram $G_v(P)$ is at the origin o, so $P \in \Theta_v \subset \Theta$.

2. If $A \in \Theta_u$, then the centroid of the parallelogram $G_u(A)$ is at the origin o. Denote its vertex by A, B, C, D anticlockwise. Then $A(x_A, y_A), B(x_B, y_B), C(x_C, y_C), D(x_D, y_D)$ belong to four quadrants, respectively. Since PA is parallel to $x-$axis, so P is situated at the second quadrant. Suppose $Q = F_v(C)$, then Q is situated at the fourth quadrant. From the facts $y_C = -y_A$, and $X_vK(y_A) = X_vK(-y_A)$, it follows that $PA = QC$. So the quadrangle $PAQC$ is a parallelogram, which centroid is the midpoint of the line segment AC, that is the origin o. So $P \in \Theta_v$, namely $P \in \Theta$.

(ii) Conversely, if $P \in \Theta = \Theta_u \cup \Theta_v$, then $F_u(P) \in$ bd$K$ and $F_u(F_u(P)) = P \in \Theta$. From the conclusion (i), we have $F_u(P) \in \Theta$. Similarly, $F_v(P) \in \Theta$ $\square$

3. Main results

Theorem 3.1. *Let K be a strictly convex body in E^2. v, u is the unit vector in directions of $x-$axis and $y-$axis, respectively. Suppose X_uK and X_vK are both even and $X_uK(0) = w_K(u)$, $X_vK(0) = w_K(v)$. Then K is an origin-symmetric convex body.*

Proof. Since $X_uK(x)$ and $X_vK(y)$ are both even, so Θ is origin- symmetric by lemma2.3. hence, it is sufficient that prove $\Theta =$ bd$K$.

Suppose $\Theta \neq$ bd$K$. From lemma 2.4, Θ is closed. So bd$K \backslash \Theta$ is made up of countable open arc segments I_α (We call I_α the component arc) and Θ is made up of countable closed arc segments J_β(We call J_β the component closed segment of Θ, it maybe degenerate into a point). Suppose

$w_K(u) = 2b$, $w_K(v) = 2a$, Then the length of $K|v^\perp$ is equal to $2b$. Since $X_vK(y)$ is even, so $K|v^\perp = [-b, b]$. Similarly, $K|u^\perp = [-a, a]$. Since K is strictly convex, so $X_uK(a)$=$X_uK(-a) = 0$. Hence the quadrangle $(a, 0), (a, 0), (-a, 0), (-a, 0)$ is a degenerated parallelogram with centroid at the origin. So $(a, 0), (-a, 0) \in \Theta$. With the same reason, $(b, 0), (-b, 0) \in \Theta$. Denote $K = \{(x, y)| - a \leq x \leq a, g(x) \leq y \leq f(x)\}$. Considering the connected component $J \subset \Theta$, which contain $(a, 0)$. It is a closed arc segment AA_4, where $A = (x_A, y_A) = (x_A, f(x_A))$. From lemma2.4, the image $F_u(AA_4)$ is also a closed arc segment of Θ. Since they both contain the point $(a, 0)$, so AA_4=$F_u(AA_4)$, $A_4 = F_u(A) = (x_A, g(x_A))$. Viewing lemma2.4 again, $A_2A_3 = F_v(AA_4)$ is also a closed arc segment of Θ, where $A_2 = F_v(A)$, $A_3 = F_v(A_4)$. Because A_2A_3 and $F_u(A_2A_3)$ both contain the point $(-a, 0)$, it follows that A_2A_3=$F_u(A_2A_3)$, and $A_3 = F_u(A_2)$. To sum up, $A_2A_3//u//AA_4$, and $AA_2//v//A_4A_3$, so the quadrangle $AA_2A_3A_4$ is a rectangle. From the facts $X_uK(x)$, $X_vK(y)$ are both even and K is strictly convex, it follows that the centroid of the rectangle $AA_2A_3A_4$ is at the origin.

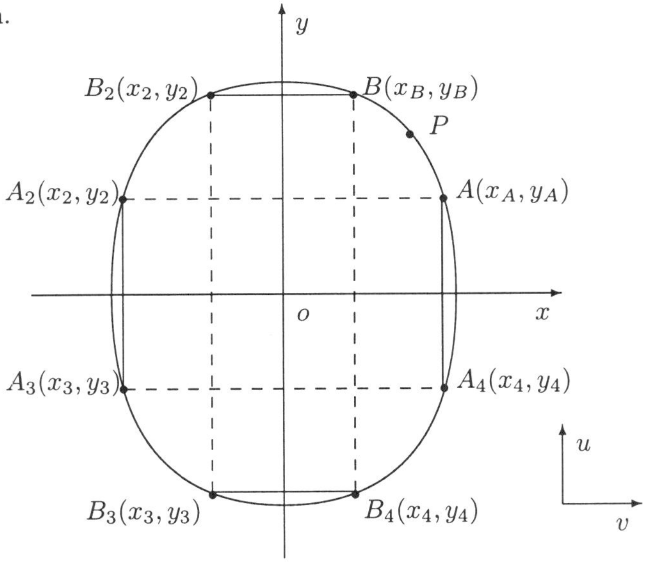

Considering the open arc segment AB of bd$K \backslash \Theta$ with A as a its endpoint, where $B = (x_B, f(x_B)$ is situated at the first quadrant. Let $B_2 = F_v(B)$, $B_4 = F_u(B)$, $B_3 = F_v(B_4)$. Using above way By lemma2.4, The quadrangle $BB_2B_3B_4$ is a rectangle with centroid at the origin. The open arcs AB, A_2B_2, A_3B_3, A_4B_4 are all component arc of bd$K \backslash \Theta$.

Suppose $L = \{(x,y)|(x,y) \in K, |x| \leq x_A, |y| \leq f(x_B)\}$. Then it can be also expressed as following:

$$L = \{(x,y)|(x,y) \in K, -x_A \leq x \leq x_A, g_L(x) \leq y \leq f_L(x)\} \tag{5}$$

where

$$f_L(x) = \begin{cases} f(x_B) & |x| \leq x_B \\ f(x) & x_B \leq |x| \leq x_A \end{cases} \tag{6}$$

$$g_L(x) = \begin{cases} g(x_B) & |x| \leq x_B \\ g(x) & x_B \leq |x| \leq x_A \end{cases} \tag{7}$$

$$= \begin{cases} -f(x_B) & |x| \leq x_B \\ g(x) & x_B \leq |x| \leq x_A \end{cases} \tag{8}$$

we can check directly that $X_uL(x), X_vL(y)$ are both even. So the centroid of the L is the origin. since open arc AB is the component arc of bd$K\backslash\Theta$, so for all $x \in (x_B, x_A)$, none of the centroids $(0, \eta_x)$ of the parallelogram $G_u(P(x, f(x))$ are at the origin. Obviously $\eta_x = \frac{1}{4}[f(x) + g(x) + f(-x) + g(x)]$ is continuous. So $\eta_x < 0$ for all $x \in (x_B, x_A)$ or $\eta_x > 0$ for all $x \in (x_B, x_A)$. Without loss of generality, Assume $\eta_x > 0$. From lemma 2.1

$$\begin{aligned} \eta_L &= \frac{1}{2\lambda_2(L)} \int_0^{x_A} X_uL(x)[f_L(x) + g_L(x) + f_L(-x) + g_L(-x)]\mathrm{d}x \\ &= \frac{1}{2\lambda_2(L)} \int_0^{x_B} X_uL(x)[f(x_B) - f(x_B) + f(-x_B) - f(-x_B)]\mathrm{d}x + \\ &\quad \frac{1}{2\lambda_2(L)} \int_{x_B}^{x_A} X_uL(x)[f_L(x) + g_L(x) + f_L(-x) + g_L(-x)]\mathrm{d}x \\ &= \frac{1}{2\lambda_2(L)} \int_{x_B}^{x_A} X_uK(x)[f(x) + g(x) + f(-x) + g(-x)]\mathrm{d}x \\ &= \frac{2}{\lambda_2(L)} \int_{x_B}^{x_A} X_uK(x)\eta_x \mathrm{d}x \\ &> 0 \end{aligned}$$

this is a contradiction with $\eta_L = 0$. So $\Theta =$bd$K$. However, Θ is origin-symmetric, so is bd$K$

□

Theorem 3.2. *Let K be a convex body with constant width in Euclidean space E^2. Suppose v, u is the unit vector in directions of $x-$axis and $y-$axis, respectively. if S_uK, S_vK are both centrally symmetric, then K is a disc*

Proof. Without loss of generality, we may assume S_uK and S_vK are both symmetric about o and let the width be a constant w. For arbitrary $P(x, y) \in S_uK$, since S_uK is origin- symmetric, we have $P'(-x, -y) \in S_uK$. Moreover, S_uK is symmetric about $x-$ axis, so $P''(-x, y) \in S_uK$. Here, $P''(-x, y)$ is the symmetric point of $P(x, y)$ about $y-$ axis, so S_uK is symmetric about $y-$ axis and therefore $X_uK(x)$ is an even. Similarly, S_vK is symmetric about $x-$ axis and $X_vK(y)$ is also an even function.

Assume the support lines of K in the direction u are L_1, L_2. And suppose $A(x, y) \in L_1 \cap \mathrm{bd}K$, $B \in L_2 \cap \mathrm{bd}K$. Then the line segment AB is parallel to $y-$ axis, otherwise the length of AB is greater than $w_K(u)$. Hence the width in the direction of AB is greater than $w_K(u)$. This is a contradiction with K is constant width.

Next we explain the point A and B both lie in $y-$ axis. Otherwise, we make the line segment $A'B'$ which is symmetric with AB about $y-$ axis. Since $X_uK(x)$ is even, so $X_uK(-x) = X_uK(x) = AB = w$. And Due to line segment $A'B'$ lie in the region between L_1 and L_2, $A'B'$ is the chord of K and therefore $A' \in \mathrm{bd}K$. Thus the width in the direction of $A'B$ is greater than (or =)the length of $A'B > w_K(u)$, which is a contradiction. Hence A, B both lie in $y-$ axe and $X_uK(0) = w$. Similarly, $X_vK(0) = w_K(v) = w$. Obviously, the convex body with constant width is strictly convex. Hence K is symmetric convex from theorem 3.1. So for arbitrary $u \in S^1$

$$h_K(u) = h_K(-u) \tag{9}$$

from the fact

$$h_K(u) + h_K(-u) = w(u) = w \tag{10}$$

it follows that

$$h_K(u) = \frac{w}{2} \tag{11}$$

is a constant, hence K is a disc. □

References

1. R.J.Gardner,*Geometric Tomgraphy* , Cambridge university press, 1995.27-29
2. R.Schneider. *Conves Bodies: The Brunn-Minkowski Theory*, Cambridge university press, 1993.37-42

THE KINEMATIC MEASURE OF A RANDOM LINE SEGMENT OF FIXED LENGTH WITHIN A TRAPEZOID

XIE FENGFAN
School of Math. and Statistics, Wuhan University
Wuhan, 430072, China
and
Department of Math., Xianning College
Xianning, 437005, China
E-mail: xiefengfanwhu@yahoo.com.cn

LI DEYI
College of Science, Wuhan University of Science and Technology
Wuhan, 430081, China
E-mail: lideyiwkd@yahoo.com.cn

This paper gives the kinematic measure of a random line segment of fixed length within an isosceles trapezoid. The results are applied to solving geometric probability problems of Buffon needle type.

1. Preliminaries and Notations

It is an interesting problem to solve the probe searching problem. Consider two orthogonal sets of parallel lines over the plane, which are assumed apart by distances e and f respectively. The lines form a lattice of rectangles. A rectangle of sides e, f $(f \leq e)$ in the lattice is called *a fundamental region.* Dropping a needle of length l $(\leq f)$ at random into the plane, we want to find the probability p of the needle N intersecting a line of the grid. This is the Laplace extension of the *Buffon needle problem.* A lattice of fundamental regions in the plane is a sequence of congruent regions $\alpha_o, \alpha_1, \cdots$, which satisfies the following conditions:

(1) Every point of the plane belongs to one and only one region α_i;

(2) Every α_i can be transformed to α_0 by a motion that takes the whole lattice onto itself.

The union of the boundaries of the regions is called *the net of the lattice.* Consider a lattice of regions that are congruent to a convex domain K. We will discuss *the generalized Buffon needle problem.* Place a segment N of length l at random in the lattice. Find the probability p that N intersects the net. The probability p is called *the hitting probability* [1]. In this paper, we want to consider a lattice of regions that are congruent to an isosceles trapezoid.

Let Q be an isosceles trapezoid in which the length of the upper base is a, the length of the lower base is b $(a < b)$, the length of hypotenuse is c and the corresponding lower base angles are $\frac{\pi}{4}$. Denote by h_1 the distance between two base sides. The distances between the two endpoints of one hypotenuse and the other hypotenuse are denoted by h_2, h_3 $(h_2 < h_3)$ respectively. The length of diagonal line of Q is denoted by d. Define P as the union of two congruent Q so that the hypotenuse overlaps, and define M as the overlapping hypotenuse. Let the left endpoint of M be the origin O; the line which M belongs to be x-axis. Let (x, y) be the midpoint of the line segment N and φ be the angle between N and the x-axis. Then every position of N intersecting with M can be determined by (x, y, φ). The 3-form $dN = dx \wedge dy \wedge d\varphi$ up to a constant factor, is the unique 3-form which is invariant under left and right translations and under inversion of the motion. It is called *the kinematic density* for the group of motions of N in the plane. By integrating dN over a domain, we obtain a measure called *the kinematic measure* of the corresponding set of the motions of N [3].

Denote by $m(\{N : N \cap M \neq \phi, N \subset P\})$ the kinematic measure of N intersecting with M and contained in the parallelogram P. Denote by $m_P(l)$ the kinematic measure of N inside a parallelogram P [2]. Denote by $m_Q(l)$ the kinematic measure of N within a trapezoid Q. Now it is clear that $m_P(l) = 2m_Q(l) + m(\{N : N \cap M \neq \phi, N \subset P\})$. In the reference [2], the authors calculated the kinematic measure $m_P(l)$. Therefore, we need to find the measure $m(\{N : N \cap M \neq \phi, N \subset P\})$. According to $m_Q(l) = \frac{1}{2}(m_P(l) - m(\{N : N \cap M \neq \phi, N \subset P\}))$, we can get explicit formulas for the kinematic measure $m_Q(l)$.

2. Main Results

For the convenience of writing, we denote as following:

$$\varphi_1 = \arctan \frac{-2l^2 + 2c\sqrt{2l^2 - c^2}}{2(c^2 - l^2)};$$

$$\varphi_2 = \arctan \frac{-2l^2 - 2c\sqrt{2l^2 - c^2}}{2(c^2 - l^2)};$$

$$\varphi_3 = \arcsin \frac{h_2}{l};$$

$$\varphi_4 = \arcsin \frac{h_3}{l};$$

$$\varphi_5 = \arccos \frac{\frac{\sqrt{2}}{2}b - c}{l}.$$

2.1. $0 < c < a < b$

$A:$ *when* $0 < c < a - \frac{\sqrt{2}}{2}a,\ h_1 < c < h_2 < h_3 < a < d < b$

$A_1:\ 0 \le l < h_1$

$$\begin{aligned} m(\{N : N \cap M \neq \phi, N \subset P\}) &= \int_0^{\frac{\pi}{4}} l \sin\varphi(c - l\cos\varphi + l\sin\varphi)d\varphi \\ &\quad + \int_{\frac{\pi}{4}}^{\pi} l \sin\varphi(c - l\sin\varphi + l\cos\varphi)d\varphi \\ &= 2cl - \frac{\pi l^2}{4} - l^2 \end{aligned} \tag{1}$$

$A_2:\ h_1 \le l < c$

$$\begin{aligned} m(\{N : N \cap M \neq \phi, N \subset P\}) &= cl(2 + \cos\varphi_1 - \cos\varphi_2) - l^2 - \frac{\pi l^2}{4} \\ &\quad + \frac{l^2}{2}(\varphi_1 - \varphi_2 + \sin^2\varphi_2 - \sin^2\varphi_1 \\ &\quad + \frac{1}{2}\sin 2\varphi_2 - \frac{1}{2}\sin 2\varphi_1) \end{aligned} \tag{2}$$

$A_3:\ c \le l < h_2$

$$\begin{aligned} m(\{N : N \cap M \neq \phi, N \subset P\}) &= cl(\cos\varphi_2 - \cos\varphi_1) - l^2 + \frac{\pi l^2}{4} \\ &\quad + \frac{l^2}{2}(\sin^2\varphi_1 + \sin^2\varphi_2 - \varphi_1 \\ &\quad - \varphi_2 + \frac{1}{2}\sin 2\varphi_1 + \frac{1}{2}\sin 2\varphi_2) \end{aligned} \tag{3}$$

$A_4:\ h_2 \le l < h_3$

$$\begin{aligned} m(\{N : N \cap M \neq \phi, N \subset P\}) &= cl(\cos\varphi_2 - \cos\varphi_3) - l^2 + \frac{\pi l^2}{4} \\ &\quad + \frac{l^2}{2}(\sin^2\varphi_2 + \sin^2\varphi_3 - \varphi_2 \\ &\quad - \varphi_3 + \frac{1}{2}\sin 2\varphi_2 + \frac{1}{2}\sin 2\varphi_3) \end{aligned} \tag{4}$$

$A_5:\ h_3 \le l < a$

$$m(\{N : N \cap M \neq \phi, N \subset P\}) = cl(\cos\varphi_2 - \cos\varphi_4) - l^2 + \frac{\pi l^2}{4} + \frac{l^2}{2}(\sin^2\varphi_2 + \sin^2\varphi_4 - \varphi_2 - \varphi_4 + \frac{1}{2}\sin 2\varphi_2 + \frac{1}{2}\sin 2\varphi_4) \qquad (5)$$

$A_6:\ a \le l < d$

$$m(\{N : N \cap M \neq \phi, N \subset P\}) = cl(\cos\varphi_2 - \cos\varphi_1) - l^2 + \frac{\pi l^2}{4} + \frac{l^2}{2}(\sin^2\varphi_1 + \sin^2\varphi_2 - \varphi_1 - \varphi_2 + \frac{1}{2}\sin 2\varphi_1 + \frac{1}{2}\sin 2\varphi_2) \qquad (6)$$

$B:$ *when* $a - \frac{\sqrt{2}}{2}a < c < \frac{\sqrt{2}}{2}a,\ h_1 < c < h_2 < a < h_3 < d < b$
$B_1:\ 0 \le l < h_1,\ A_1;$
$B_2:\ h_1 \le l < c,\ A_2;$
$B_3:\ c \le l < h_2,\ A_3;$
$B_4:\ h_2 \le l < a,\ A_3;$
$B_5:\ a \le l < h_3,\ A_6;$
$B_6:\ h_3 \le l < d,\ A_6;$
$C:$ *when* $\frac{\sqrt{2}}{2}a < c < a,\ h_1 < h_2 < c < a < h_3 < d < b$
$C_1:\ 0 \le l < h_1,\ A_1;$
$C_2:\ h_1 \le l < h_2,\ A_1;$
$C_3:\ h_2 \le l < c,\ A_2;$
$C_4:\ c \le l < a,\ A_3;$
$C_5:\ a \le l < h_3,\ B_5;$
$C_6:\ h_3 \le l < d,\ B_6;$

2.2. $0 < a < c < b$

$D:$ *when* $0 < a < \frac{\sqrt{2}}{2}c,\ h_2 < a < h_1 < c < h_3 < d < b$
$D_1:\ 0 \le l < h_2,\ A_1;$
$D_2:\ h_2 \le l < a,\ A_1;$

$D_3: \ a \le l < h_1$

$$m(\{N : N \cap M \neq \phi, N \subset P\}) = cl(2 + \frac{\sqrt{2}}{2} - \cos\varphi_1) - \frac{1}{2}bl - \frac{\pi l^2}{2} + \frac{\sqrt{2}}{2}bl\cos\varphi_3 - \frac{l^2}{2}(\sin^2\varphi_1 - \varphi_1 - \varphi_3) - \frac{l^2}{4}(1 + \sin 2\varphi_1 + \sin 2\varphi_3) \quad (7)$$

$D_4: \ h_1 \le l < c, \ A_2;$
$D_5: \ c \le l < h_3$

$$m(\{N : N \cap M \neq \phi, N \subset P\}) = \frac{\sqrt{2}}{2}bl(\cos\varphi_2 - \cos\varphi_5) - \frac{\pi l^2}{4} + \frac{l^2}{2} - \frac{l^2}{4}\sin 2\varphi_2 + cl\cos\varphi_5 + \frac{l^2}{2}(\varphi_2 - \sin^2\varphi_5) \quad (8)$$

$D_6: \ h_3 \le l < d, \ B_6;$
$E: \ when \ \frac{\sqrt{2}}{2}c < a < c, \ h_2 < h_1 < a < c < h_3 < d < b$
$E_1: \ 0 \le l < h_2, \ A_1;$
$E_2: \ h_2 \le l < h_1, \ A_1;$
$E_3: \ h_1 \le l < a, \ A_2;$
$E_4: \ a \le l < c, \ A_2;$
$E_5: \ c \le l < h_3, \ D_5;$
$E_6: \ h_3 \le l < d, \ D_6;$

3. Application of $m_Q(l)$ to geometric probability

Consider two sets of parallel lines over the plane, the angle between two is $\frac{\pi}{4}$, which are assumed apart by distances $a+b$, c respectively. The lines form a lattice of parallelogram. A parallelogram of sides $a+b$, c the angle between two is $\frac{\pi}{4}$ in the lattice is called *a fundamental region*. For every parallelogram, we use a segment M of length c connecting two parallelling sides of length $a+b$. Then M divide every parallelogram into two congruent isosceles trapezoids. A trapezoid in which the length of the two base are a, b $(a < b)$, the length of two hypotenuses are c and the two base angles are $\frac{\pi}{4}$ in the lattice is called *a fundamental region.* That is to say that the motion that superimposes on every α_h ,α_s may be translation or rotation. Dropping a needle N of length l at random into the trapezoid lattice, we want to find the probability p of the needle N intersecting with a line of the grid. If F denotes the area of Q and $m_Q(l)$ denotes the kinematic measure of N, from the reference[1]

$$p = 1 - \frac{m_Q(l)}{\pi F} = 1 - \frac{2\sqrt{2}m_Q(l)}{\pi(a+b)c}$$

According to the the individual expressions for the kinematic measure $m_Q(l)$, we can get the solution to Buffon problem in different conditions. For example, due to $m_Q(l) = \frac{1}{2}(m_P(l) - m(\{N : N \cap M \neq \phi, N \subset P\}))$, when $A_1 :\ 0 \leq l < h_1$:

$$\begin{aligned} m_P(l) &= \frac{\sqrt{2}}{2}\pi(a+b)c - 2[(a+b+c)l - \frac{\pi l^2}{8} - \frac{l^2}{2}] \\ &= \frac{\sqrt{2}}{2}\pi(a+b)c - 2(a+b+c)l + \frac{\pi l^2}{4} + l^2 \qquad (9) \\ m_Q(l) &= \frac{1}{2}[\frac{\sqrt{2}}{2}\pi(a+b)c - 2(a+b+c)l + \frac{\pi l^2}{4} + l^2 \\ &\quad -(2cl - l^2 - \frac{\pi l^2}{4})] \\ &= \frac{1}{2}[\frac{\sqrt{2}}{2}\pi(a+b)c - 2(a+b+2c)l + \frac{\pi l^2}{2} + 2l^2] \qquad (10) \\ p &= 1 - \frac{\sqrt{2}[\frac{\sqrt{2}}{2}\pi(a+b)c - 2(a+b+2c)l + \frac{\pi l^2}{2} + 2l^2]}{\pi(a+b)c} \\ &= \frac{2\sqrt{2}}{\pi(a+b)c}[(a+b+2c)l - \frac{1}{4}\pi l^2 - l^2] \qquad (11) \end{aligned}$$

References

1. Ren Delin, *Topics in Integral Geometry, World Scientic: Singapore, New Jersey, London, and Hongkong,* (1994).
2. Zhang Gaoyong, Li Rongze, *Wuhan Iron and Steel University* **1:1**, 106-128 (1984).
3. L.A. Santalo, *Integral Geometry and Geometric Probability, Addison-Wesley Publishing Company,* (1976).